Readers will be enthralled by Denis Lamoureux's compelling personal story and his compelling evidence. His strong Christian testimony and his conviction that evolution is our best way to understand God's creative work can be life-changing for those who feel trapped in the belief that they have to choose between creation and evolution. The evidence says they don't, and Lamoureux's clear presentation tells them why. Even those who are not inclined to accept an evolutionary model for creation will find a new appreciation for how committed Christians who take the Bible seriously can accept such a model and yet maintain the integrity of their faith.

—**JOHN H. WALTON,** Professor of Old Testament, Wheaton College

In clear and engaging fashion, Dr. Lamoureux weaves together the tales of his own spiritual and intell wrestlings, showing how "evolutionary creation" enables him to b tually fulfilled theist." I especially appreciate his recur fallacious either/or thinking that we trap ourselves w or not (or a little of both!), you will find Lamoureux a partner.

") **COLLINS,** Professor of Old enant Theological Seminary

Writing in a captiv Denis Lamoureux shares his personal journey from Ch nd back to Christianity and from atheistic evolution to tionary creation. The book radiates his love and passion for Jesus arly explaining the evidence he learned while earning three doctoral degrees in dentistry, theology, and evolutionary biology. This book should be required reading for every scientist and non-scientist interested in science and Christian faith.

—**RANDY ISAAC,** Executive Director Emeritus, American Scientific Affiliation and Retired IBM Research Vice-President for Science and Technology

The singular contribution of this book is the emphasis on the problem in the church and in the science laboratory on either-or thinking, and Denis Lamoureux proves this is a false and dangerous dichotomy. In *Evolution: Scripture and Nature Say Yes!*, Lamoureux encourages Christians to listen to the Bible and to science. Honest, fair, and reasoned listening leads to an explosion of joy and freedom as one sees in both the Bible and nature the glory of our God. I pray that Christians across the spectrum will give this book a fair hearing. If they do, we will all grow.

—Scot McKnight, Julius R. Mantey Professor of
New Testament, Northern Seminary

Denis Lamoureux's elegant biblical and biological scholarship is presented at a level accessible to anyone and illuminates a path where Christians can engage both the Word of God and biological science in harmonious dialogue. Lamoureux is intimately aware of the struggles faced when we look for peace between the evidence for evolution in the natural revelation of God and the Truth in the Holy Scriptures. Trained to the doctoral level in both theology and biology, he utilizes his own journey through these struggles to provide a pastoral and accommodating exposition of the intersection between the ancient science presented by the Holy Spirit-inspired biblical writers and the compelling evidence of a God-ordained and sustained process of evolution.

—Nancy Halliday, PhD, Associate Professor of
Cell Biology, University of Oklahoma College of
Medicine

This is a successful and effective academic project, but more than this, it is an utterly convincing decades-long autobiographical account of God's faithful revelation through the author's attentive reading of both the *Book of God's Word*, written in Scripture, and the *Book of God's Works*, written in the material creation that science is privileged to consider.

While arguing for an evolutionary dimension to God's creation, the author is concerned to refute damaging skepticism towards scientific insight, on the part of theologically conservative North American Christians. The outcome provides a more vigorous witness to a scientifically savvy world . . . which includes both young people being raised in Christian communities, and unbelievers for whom the Church seeks an entrée for evangelism!

—Ian Johnston, Professor of Biological Sciences,
Bethel University

No one is more qualified to write about evolution and Christian faith than Denis Lamoureux, who has studied both areas extensively for many years. I recommend this book not only for the wealth of important information it contains, but especially for it's wonderful attitude of honest inquiry coupled with genuine zeal for the future of the body of Christ. Written by a teacher and scholar with pastoral sensitivity, it brings the voice of personal experience to bear on an issue that challenges many young people and documents the damage done to the next generation by some promoters of young-earth creationism.

—**EDWARD B. DAVIS,** Professor of the History of
Science, Messiah College

Please listen carefully to Denis Lamoureux's fascinating story. He struggled a lot and finally found peace between scientific integrity and his true love for Jesus. Reading this book will save you so much time and trouble, giving you the precious clues that will enable you to escape from the confusing maze of the origins debate.

—**BENOIT HÉBERT,** President of *Science et Foi*
(Science and Faith), www.scienceetfoi.com

Denis O. Lamoureux is passionate about integrating science and a biblical-based faith in regards to the origins and evolution discussion. He writes in an understandable fashion for the layperson and scientist alike, clearly explaining the scientific and theological rationale for the evolutionary creationist perspective while maintaining a high regard for biblical inerrancy and scientific information.

His love for Christ and God's created world is a common theme throughout the book, and his constant desire is to enable all believers, young and old alike, to come to a reasoned and God-inspired perspective on the origin of the biological diversity. Denis weaves his own journey to faith and reconciliation with science throughout the book in such a way as to serve as an example to all who are interested in this discussion.

I have interacted with Denis personally and on seminar panels, and I can assure readers that he loves God and God's people and does not denigrate any believer or their views on this hotly debated topic. Denis uniquely balances grace and tenacity in supporting his position. Anyone interested in the creation-evolution discussion needs to seriously engage with Dr. Lamoureux's evolutionary creationist perspective of the creative work of God in His world.

—**HAL C. REED,** PhD, Professor of Biology,
Department of Chemistry and Biology, Oral
Roberts University

Every year students in high school and college who come from Bible-believing churches are faced with the dilemma, "Do I give up my faith and listen to what the science teachers say, or do I reject science and keep my faith?" This scenario has played out for all too many and for all too many years! Now Dr. Lamoureux, like a "Catcher in the Rye," uses his advanced degree studies in theology and biology to give students and the church a way to keep their faith, see even more glorious things in Genesis 1, and yet grasp all that the best science has to offer.

I recommend this book to Intervarsity staff (my former co-workers in the academic world), pastors, parents, and students who are currently struggling with the "literal" interpretation of Genesis and what the science texts and professors are saying. Carefully working through this book will help strengthen faith and open otherwise closed gateways to the exciting world of God's creation as science is coming to understand it.

—**TERRY MORRISON,** PhD, Emeritus Director,
Faculty Ministry, Intervarsity Christian Fellowship

This book is a magnificent contribution to Evangelicalism. It is a searchlight cutting through the fog of the Bible vs. science debate. Lamoureux shows how evolution and all of science can be integrated with a full commitment to Jesus Christ and to the divine authority of Scripture. Its crowning benefit is that every chapter breathes out Jesus!

—**PAUL H. SEELY,** author of *Inerrant Wisdom: Science and Inerrancy in Biblical Perspective*

In *Evolution: Scripture and Nature Say Yes!,* Dr. Denis Lamoureux uses his personal experiences to powerfully illustrate why so many Christian students lose their faith in college over questions of faith and science. He then provides a remarkably accessible and comprehensive outline of the various assumptions that Christians make regarding how they interpret Scripture and understand natural history, opting for the view that the science in Scripture reflects the ancient understanding of nature at the time the Bible was written. Finally, he argues persuasively that remarkable evolutionary pathways point to an intelligently designed universe and uses the writings of both Charles Darwin and Richard Dawkins to show how they struggled with the clear evidence of design in nature.

—**WALTER L. BRADLEY,** PhD, Emeritus Professor
of Mechanical Engineering and Emeritus
Distinguished Professor of Mechanical Engineering,
Texas A&M University and Baylor University

Anyone who has ever wondered about the relationship between Christian faith and modern science should read this book. It makes a crystal clear and highly compelling case for calling a halt to the alleged warfare between Christianity and evolution. Denis Lamoureux's unique personal story as a theologian-scientist, combined with careful scholarly argumentation and unashamed faith in Jesus, makes this book very hard to put down. It will be an oasis in the desert for many readers. Unlike so many other writings on science and religion, this book is generous toward other views and invites readers to draw their own conclusions.

—JEFFREY P. GREENMAN, President, Regent College

This is a great book to hand to a Christian student or friend who is asking questions about evolution. Denis Lamoureux, a pioneer of Christian engagement with evolution as both a theologian and a scientist, roots his arguments in the classic doctrine of the two books—God's word and God's world. But even more than that, his pastoral heart comes through in these pages as he shares his own story. Denis shows that people don't have to choose between Scripture and nature but can say "Yes!" to both.

—DEBORAH HAARSMA, President of BioLogos

EVOLUTION:
SCRIPTURE AND
NATURE SAY YES!

EVOLUTION:
SCRIPTURE AND NATURE SAY
YES!

DENIS O. LAMOUREUX

ZONDERVAN

Evolution: Scripture and Nature Say Yes
Copyright © 2016 by Denis O. Lamoureux

This title is also available as a Zondervan ebook.

Requests for information should be addressed to:
Zondervan, *3900 Sparks Drive SE, Grand Rapids, Michigan 49546*

ISBN 978-0-310-52644-5

All Scripture quotations, unless otherwise indicated, are taken from The Holy
Bible, New International Version®, NIV®. Copyright © 1973, 1978, 1984, 2011 by
Biblica, Inc.® Used by permission of Zondervan. All rights reserved worldwide.
www.Zondervan.com. The "NIV" and "New International Version" are trademarks
registered in the United States Patent and Trademark Office by Biblica, Inc.®

Scripture quotations marked NRSV are taken from the *New Revised Standard Version
of the Bible.* Copyright © 1989 National Council of Churches of Christ in the United
States of America. Used by permission. All rights reserved.

Scripture quotations marked NKJV are taken from the New King James Version®.
Copyright © 1982 by Thomas Nelson. Used by permission. All rights reserved.

Any Internet addresses (websites, blogs, etc.) and telephone numbers in this book
are offered as a resource. They are not intended in any way to be or imply an
endorsement by Zondervan, nor does Zondervan vouch for the content of these sites
and numbers for the life of this book.

All rights reserved. No part of this publication may be reproduced, stored in
a retrieval system, or transmitted in any form or by any means—electronic,
mechanical, photocopy, recording, or any other—except for brief quotations in
printed reviews, without the prior permission of the publisher.

Cover design: John Hamilton Design
Cover image: Shutterstock.com
Interior design: Denise Froehlich

Printed in the United States of America

17 18 19 20 /DCI/ 18 17 16 15 14 13 12 11 10 9 8 7 6 5 4 3 2

Dedicated to Professor Loren Wilkinson,
who challenged me:

"Denis, if you gave up your belief in six day creation,
would you also give up your faith in Jesus?"

CONTENTS

ACKNOWLEDGMENTS

I am so grateful for many people who have assisted and encouraged me during the writing of this book. My teaching assistant Anna-Lisa Ptolemy worked tirelessly and meticulously on editing the manuscript. Many thanks to her husband, Scott, and her children—Isabel, Nathaniel, and Jacob—for sharing her time during this writing project. Madison Trammel at Zondervan suggested invaluable insights regarding the structure and tone of the book. Jim Ruark at Zondervan has been a source of strength over the years. Thank you, Jim, for making me a better writer. And President Terence Kersch at St. Joseph's College in the University of Alberta has thoroughly supported me in the writing of this book.

Many thanks to Andrea Dmytrash, John Hamilton, Kenneth Kully, Michael Caldwell, and Braden Barr for their artwork. Others who have contributed include Grace Barlow, Chris Barrigar, Lyn Berg, Daneel Blair, Jeremy Cooper, Mona-Lee Feehan, Denise Froehlich, Keith Furman, Brian Glubish, Wendell Grout, Sarah Gombis, Jane Haradine, Steve Hewko, Randy Isaac, Douglas Jacoby, Mark Kalthoff, Nathan Kroeze, Bob Lamoureux, Don McLeod, Sara McKeon, Don Robinson, Anita and Paul Seely, Jennifer Swainson, Kevin Tam, Jon Thomas, and Danica Wolitski.

Finally, I want to thank Ozzie and Bernice Lamoureux, better known as Mom and Dad, who provided a loving home and a healthy respect for education. Our family meal every Sunday evening has been such a blessing over the years in revealing the Lord's incarnational love.

Our Saviour Jesus said, "You are in error because you do not know the Scriptures or the power of God" (Matt. 22:29). He has placed before us two books to study so that we will be protected from error. First is the Book of Scripture, which reveals the will of God; and second is the Book of Nature, which expresses his power. The latter is a key to the former. Not only does it open our understanding to conceive the true sense of the Scriptures . . . but it primarily opens our belief, in drawing us into a meditation of God's almighty power, which is predominantly signed and engraved upon his works.

To conclude, let no one . . . think or maintain that anyone can search too far or be too well studied in the Book of God's Words or in the Book of God's Works—theology or science. But rather, let everyone endeavor an endless progress or proficiency in both.

—SIR FRANCIS BACON (1561–1626)[1]

CHAPTER 1

TRAPPED IN "EITHER/ OR" THINKING

"I am so mad!" These words were cried out by a young woman after she stood up during the middle of a class in my college course on science and religion. I slowly sank down behind the lecture stand, quickly glancing at my notes, hoping to find what I had said that triggered her outburst. And she wasn't finished. "I'm mad at my parents for putting me in an expensive Christian school where teachers taught me that Satan had concocted the so-called 'theory' of evolution. I am angry with my youth pastor for telling me that I had to choose between evolution and creation. And I'm furious with the senior pastor at my church. On Sunday mornings, he has preached that evolutionists cannot be true Christians!"

Finally, she sat down. I relaxed, realizing that her rage was not directed at me. I said to her, "You must always remember that your parents love you and care for you. The fact that they paid a lot of money to send you to a private Christian school instead of public school shows they were giving you the opportunity for your faith to mature. I am also certain that the pastors in your church wanted the best for you. But many of them are taught in theology school that evolution can destroy our faith in Jesus."

I added, "I completely understand your frustration. I have lived your story of being forced into making a choice between evolution and creation, and this led me to pick science over my Christian faith. But as you are seeing in this course, there are a variety of different ways to embrace both God and modern science. My challenge to you and your classmates is this: You are the next generation of leaders in the church. What can you do to improve this situation?"

Like many Christians today, the parents and pastors of my students are fearful of evolution. They assume that if evolution is true, then God did not create the world and Christianity must be false. More specifically, if the universe and life were not made in six literal days about six thousand years ago, then the Bible is a lie, and we cannot trust anything it says about Jesus and our salvation. Some people go so far as to say that if life evolved, then God does not exist. I understand this type of thinking because I lost my faith as a freshman in college after taking just one course on evolution. I was trapped in the assumption that I had to choose between evolution and creation, and between modern science and Christian faith. By the time I left college, I was an atheist.

As an evolutionist, I assumed that the entire cosmos and every living organism had evolved only through blind chance. My existence had no purpose or ultimate meaning whatsoever. With such a dark and dreary picture of the world, I thought that the best way to live my life was to live for me and me alone. Selfishness marked my lifestyle. I didn't believe in morality because I didn't think there was any ultimate right or wrong. From my perspective, humans were nothing but animals evolved through a completely random process. I saw myself to be no different than any other animal and tried living like one. I'll spare you the details. I'm embarrassed by things I did and people I hurt. For many Christians, this is another fear of evolution. They assume that it leads to immorality.

My story is not unique. Many Christian students have lost their faith over the issue of evolution. Today they are leaving Christianity in greater numbers and at a much earlier age. A 2011 Barna Group survey reveals that 59 percent of young people "disconnect either permanently or for an extended period of time from church life after age 15." The issue of science is one of the reasons why they leave the church. This study records that 25 percent of them perceive that "Christianity is anti-science," and 23 percent have "been turned off by the evolution-versus-creation debate."[1] Every Christian should be concerned with this shocking loss of faith among the younger generation.

I wrote this book to address problems Christian students face regarding modern science and the origin of the world. It is based on the traditional belief of a fruitful relationship between God's Two Books. The Book of God's Words is the Bible. Scripture reveals spiritual truths concerning our Creator, his creation, and us. The Book of God's Works is the physical world. Science is a wonderful gift from the Lord to explore the heavens and the earth and all the wonderful creatures. Through microscopes and telescopes we can see that nature "declare[s] the glory of God" (Ps. 19:1). Together, God's Two Books—Scripture and nature—provide us with a divine revelation of *who* created the world and *how* he created it.

My hope for this book is that it helps you to develop your view of the relationship between Christianity and science and in particular your position on origins. My goal is for readers to have a stronger God-centered biblical faith and a great admiration for modern science, including evolution. It's worth pointing out that there is not just one way to view God's Two Books, but many different ways. I will also share with you my own approach to relating Scripture and nature.

This book is personal. It's not merely about theories and concepts. It's about an experience I had. I wrestled with evolution and Christian faith over a period of about twenty years. I'll share many mistakes that I made, and hopefully you will avoid making these same errors.

Forced to Choose between Evolution and Creation

When the topic of origins first came to my attention, there was a serious problem with the way I thought about evolution and Christianity. Like most people, I was trapped in "either/or" thinking. This forced me to assume that there are only two possible positions with regard to the origin of the universe and life. I thought I had to choose *either* evolution *or* creation.

An issue that is divided into only two simple positions is called a

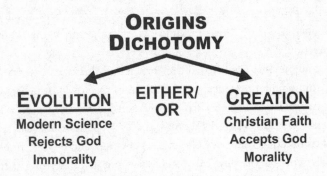

Figure 1–1. The Origins Dichotomy.

"dichotomy." This term is made up of the Greek word *dicha*, which means "in two," and *temnō* is the verb "to cut." A dichotomy forces people to pick between one of two choices. It completely blinds them from seeing a wide range of other possibilities. For example, why could there not be a third choice for origins, where the God of Christianity used evolution as his method to create the entire cosmos and every plant and animal?

Figure 1–1 presents the origins dichotomy. It has been incredibly destructive both inside and outside of the church. For those who love modern science and accept evolution, this dichotomy has driven them away from God and even led to immoral lifestyles. It has also forced a number of scientists to assume that in order to become a Christian, they have to give up their scientific views. On the other hand, many Christians reject evolution and modern science because the origins dichotomy is taught in their churches. Evolution is presented as a lie that Satan uses to destroy belief in God. This fuels fear and turns Christian students away from studying science, especially the age of the earth and the evolution of life.

Personally, I know the destructive power of the evolution vs. creation dichotomy. I was raised in a good Christian home and went to a fine Christian school. During a biology class in high school, I was introduced to the theory of evolution. One of my favorite teachers pointed out that evolution could be seen as God's way of creating all living organisms. This teacher, a wonderful Christian, added that biological evolution was not a reason for us to reject our faith.

However, looking back at this critical moment in my life, I now realize that just *telling* high school students that God could have used evolution is not enough information to protect them from the origins dichotomy. Students need to be given a variety of reasons why it might be true that the Lord created through an evolutionary process. Over the years I have spoken to high school students on many occasions. They always ask excellent questions about origins. Parents, pastors, and Christian teachers need to deal directly with these questions. They have to stop giving routine shallow answers that are entrenched in the origins dichotomy.

When I left high school, I was not equipped to protect my faith from the attacks of a secular education at a public college. Within only the first few weeks of my freshman year, professors and older students were telling me that intelligent people do not believe in God. According to them, religion is just a phase of human evolution, and we are in the process of shedding it away. They added that the only people who accept God are those who are either uneducated or near their death.

My college education was also claiming that science is the only credible form of thinking because science deals with hard facts and real truth. After all, just look around at how our lives have been improved by amazing scientific discoveries. NASA, not religion, put humans on the moon. Medicine, not prayer, cures diseases. In this way, I was being trapped in another notorious dichotomy. College was forcing me to choose *either* science *or* religion.

My first biology course in the first term of college was on evolution. The basic message was quite obvious. Life originated through an evolutionary process with no plan, no purpose, and definitely no God. Humans were nothing but an unintended fluke of nature. I got the distinct impression that to be a real scientist, you had to be an atheist. By the end of the course, I came to what I thought was a completely logical conclusion: since evolution is true, then the Bible must be false and Christianity a lie.

Like most Christians, I read the Bible literally, and I knew that the first chapter of the book of Genesis stated the world was created in six 24-hour days. However, in biology class I was being shown a wide range of scientific evidence for the evolution of living organisms over billions of years. I was hopelessly trapped in the origins dichotomy and "either/or" thinking. I had no idea that there was an intellectually respectable middle position between atheistic evolution and six day creation.

Losing My Faith in College

Toward the end of my first term, my parents discovered that I was no longer attending church. At the Christmas break, they sat me down to find out why. I can remember the conversation as if it were yesterday. Actually, it wasn't a conversation. It was a nasty argument. I pounded on the kitchen table, raised my voice, and claimed that evolution is a fact and that the world was not created in six days. To the horror of my parents, I then told them that the Bible is complete nonsense and nothing but a fairy tale. Needless to say, this was not the Christmas present they were expecting.

My mother and father did not have the privilege of a college education. They could not respond to my scientific arguments. In fact, I was the first in my family to go to college. There was no aunt, uncle, or cousin who could have helped me with questions about evolution and Christianity. As I mentioned earlier, I simply wasn't equipped to face the attacks of a secular education on my faith.

This is the main reason why I have written this book. I want Christian students to be prepared to face the challenges of secular education. I want them to know all their options so they can make informed decisions about what they believe about science and faith. I don't want students to get trapped in the origins dichotomy or the science vs. religion dichotomy. This is the book I wish I could have read when I entered college.

At the end of my first year of college, I wrote in my diary, "It seems that humans are nothing but mere chemical reactions programmed by

DNA. . . . But there's more, I'm sure." Although I had rejected the God of Christianity, I did not at that time become an atheist. I still believed in a god who was responsible for humans being "more" than just "mere chemical reactions." I was sort of spiritual, but I was definitely not religious. I thought that religion was a scam to make money and that most ministers and priests were crooked.

In reality, I lived as if this god did not exist—except when I desperately needed him, such as when I thought my girlfriend was pregnant. In other words, I had a so-called "god-of-the-emergencies." I prayed to him only to save me from my stupidity and immorality. Clearly, I was a hypocrite.

After two years of college, I was accepted into dental school. I joined the military to pay for my education. A number of my classmates were marvelous Christians. They often shared their faith with me. I even found some of their reasons for believing in Jesus to be persuasive. But it was their godly lifestyle that impacted me more than any argument for religion. I wanted what they had even though I couldn't put it into words at the time. I wanted holiness, and I wanted God in my life. Yet the issue of origins was the looming problem. My Christian friends rejected evolution. So once again, that dreadful dichotomy appeared in my life and forced me to assume I had only two choices: *either* purposeless evolution with no God *or* creation in six 24-hour days through the God of Christianity.

Although the powerful reality of Jesus was being brilliantly displayed in the lives of some of my dental classmates, I gradually moved away from believing in a mostly absent god during my early years of college. I slipped in and out of periods of uncertainty regarding whether God existed (called "agnosticism") until finally I became an atheist.

In another revealing entry in my diary I wrote, "Love is a protective response characteristic of all animals, except expressed to greater levels in humans because of their superior intelligence." From my perspective, love was merely an illusion. It was just a meaningless idea found in romance novels and fairy tales that weak-minded people had invented. What everyone called "love" was nothing but an animal instinct selected by evolution so that living organisms could continue to reproduce and evolve. With this view of love, it takes little effort to imagine how I treated women. I'm not proud of that. For me, an atheistic view of evolution led to an immoral lifestyle.

I graduated from dental school and began to serve as a military dentist. My secular college education had completely brainwashed me into believing that happiness was to be found in a self-serving lifestyle. This meant treating women disrespectfully, abusing drugs and alcohol, driving fast cars, and playing sports as much as possible. I partied as hard as I could. From a distance, it looked like I was having the time of my life. But inwardly, I sensed that there was something wrong with my selfish way of living. At a logical level, I recognized the foolishness in running around, getting high, buying cars, and being consumed by sports. Deep inside me, I had an uncomfortable feeling. There was a profound emptiness in my soul. I also had a distinct sense of being dirty and unclean.

Returning to Faith, but Still Trapped in Dichotomies

One of my first military posts was on the island of Cyprus in the Mediterranean, where I served as a United Nations peacekeeper. By God's grace and in an answer to my mother's prayers, it was during that six-month tour that I returned to my Christian faith through reading the gospel of John in the Bible. There were no dramatic events or major crises in my life at that time. I was completely fed up with living a selfish and filthy lifestyle. I wanted God in my life and I wanted holiness. As I went through the gospel of John, I could sense the Lord gently working in my heart. And there was a feeling of being freed and cleansed from the dirt of my sins.

On Good Friday, I went to church for the first time in more than seven years. God revealed to me the meaning of his death on the Cross: Jesus loves us so much that he died for us (John 3:16). Think about that. The Creator of this astonishing universe suffered death on the Cross to show how much he loves each and every one of us! I began to weep during the Scripture reading on the Crucifixion and continued for the rest of the service. Miraculously, the Lord's peace entered my soul.

The Cyprus posting changed my life forever. I have never once regretted my return to Christianity. When I left home to join the United Nations peacekeepers, I was spiritually empty. I returned filled with the peace of Jesus. Indeed, I had been born again (John 3:3). Thanks to God's love, a military peacekeeper met the Prince of Peace.

During my Cyprus tour of duty, I went to Israel for a holiday. In a dark and dusty little bookstore in Tel Aviv, I stumbled upon Duane Gish's famous book *Evolution: The Fossils Say No!* Gish was one of the most important anti-evolutionists in the world at that time. Filled with excitement, I went directly to the beach and read his 120-page book in just one afternoon. It rocked my world. Gish thoroughly convinced me that a massive conspiracy was happening within the scientific community. He argued that there was no fossil evidence to support evolution and that this fact was being withheld from the public.

Soon after returning from Cyprus, I became a member of a church. The pastor was wonderful, and he always based his preaching on the Bible. It was great to be surrounded by young adults my age who loved God and wanted to live a holy life. Even though I was now a recommitted Christian, I was still trapped in "either/or" thinking and entrenched in the origins dichotomy.

In fact, most of the members of my church were as well. They despised evolution and argued intensely that the world was created in only six days about six thousand years ago. They also told me that the godly way to read the creation account in Genesis 1 was to read it literally. My church convinced me that evolution was the Devil's primary weapon for attacking the faith of young people. This made perfect sense to me because, as a college freshman, it took just one course on evolution to destroy my Christian faith.

During this time, Sunday school teachers warned me about so-called "theistic evolution." (The Greek word *theos* means "God.") This view of origins claims God created the universe and life through evolution. Of course, I certainly did not read that in the Bible! Genesis 1 clearly states that the Creator made the entire world and all plants and animals in just six 24-hour days. My teachers also told me that theistic evolution is accepted only by so-called "liberal Christians." I was given the distinct impression that theistic evolutionists are not Christians because they aren't really committed to Jesus and they don't fully trust the Bible. Attending Sunday school led me to believe that *true* Christians are six day creationists.

I became totally consumed by the topic of origins and bought the best books defending creationism. I went to seminars to learn as much as I could about this view of origins. Every time there was an evolution vs. creation debate in my city, I was there. I became friends with one of the most important six day creationists in the nation. I also attended a summer school sponsored by the Institute for Creation Research (ICR) in El Cajon, California.[2] This is a Christian organization of scientists who work in a subject area called "creation science." Their main theory is that creation in six days can be proven scientifically. I was hooked and soon became an ICR supporter.

The work of creation scientists inspired me to write an article in the journal *Creation Science Dialogue* entitled "Philosophy versus Science." I eagerly promoted my newly found anti-evolutionary views and ended the article by stating, "I challenge anyone who takes pride in their objectivity to entertain seriously scientific creationism. It may very well be the most important study of your life."[3]

I am sure you have identified the problem with my article. As the title reveals, I was still trapped in the origins dichotomy. I did not believe that evolution was science, but merely a secular "philosophy." Real "science" was in fact creation science. My "either/or" way of thinking forced me into believing that a secular conspiracy was happening in colleges and universities throughout the nation. I didn't think there was any scientific evidence whatsoever to support evolution.

A fire started to burn in my soul. I wanted to become a creation scientist to attack evolutionists for brainwashing college students with

Satan's lie that the world had evolved over billions of years. The more I read about creation science, the more convinced I became that it was impossible for a *real* Christian to be an evolutionist. And I came to the conclusion that *the* Christian position on origins had to be creation in six literal days about six thousand years ago.

But again I ask the question I posed earlier: Why couldn't the God of Christianity have used evolution as his method to create the universe and life? And a related question: Is there a godly way to read the account of creation in Genesis 1 that isn't literal? In other words, is there an approach that honors the Bible as being truly the inspired Word of God and focuses on the truths that the Lord intended for us?

CHAPTER 2

OPENING GOD'S TWO BOOKS

As a recommitted and maturing Christian, I wrestled with the issue of God's will. Like many young people, I wondered if his will for my life was simply to follow the Ten Commandments and the moral laws in the Bible. Or does the Lord call us to do something specific? After a lot of heart-wrenching struggles, I slowly came to the belief that God wants *both*. He wants us to live holy lives, and he does have a wonderful plan for each of us. After all, he created us and knows us better than we know ourselves! The Lord loves us and knows exactly what calling will provide the most joyful and purposeful life imaginable.

As I came to the end of my military duty, a number of amazing opportunities arose. My commanding officer sent me to work with a jaw surgeon in a hospital to see if I was interested in that dental specialty. The military even offered to pay my entire dentist's salary over five years if I wanted to go to medical school. There was also the prospect of practicing dentistry in a lucrative office. Although I was blessed to have these amazing possibilities, I battled with what to do with the next stage of my life. I was terribly unhappy and quite depressed.

In the middle of my most intense struggles, the Lord sent a mature Christian who said to me, "Denis, when you are on your knees before God, and your heart is open, you will know what he wants you to do with your life." In those quiet moments on my knees, I knew in my heart of hearts my calling and exactly what he wanted me to do. But did I listen to the Lord's call?

No. I went to medical school and lasted a grand total of three days! It was my Jonah experience. God had called Jonah to preach in the city of Nineveh, but he "ran away from the LORD and headed for Tarshish" (Jonah 1:3). Don't get me wrong—medicine is a fine career if you are called by God to be a doctor. But I wasn't. In those precious moments on my knees with an open heart, it was clear that the Lord was calling me to become a creation scientist in order to attack evolutionists in secular colleges and universities. My faith had been destroyed by one first-year college course on evolution, and I wanted to protect students from Satan's lie that life had evolved.

To equip myself for the battle, I decided to get a PhD in theology and a PhD in biology. First, I planned to study the Book of God's Words with a focus on the creation accounts in the Bible. Afterward, I would study the Book of God's Works, beginning in a creation science program at the Institute for Creation Research. I believed that by becoming both a theologian and a biologist, I would be well-prepared to "fight the good fight" (1 Tim. 1:18 NRSV) against the Devil and his evolutionist disciples.

Opening the Book of God's Words

I had the privilege of attending one of the best theology schools in the nation. My research focused on the accounts of origins in Genesis 1–11. But I had a much larger agenda, as my diary records on the first day of school: "The Grand Plan: Declare absolute & pure hell on the 'theory.'" Like most Christians, I assumed that the so-called "theory" of evolution was nothing but nonsense.

However, "The Grand Plan" soon came under attack. Like all theology students, I discovered that interpreting the Bible is more complicated than what we learn in Sunday school. Only weeks into my first term, one of the world's greatest theologians stated in class that "the biblical creation accounts were obviously written in picture language." I knew that this professor was a marvelous Christian. Many of his books were very helpful in my walk of faith. I had even met people who came to the Lord through his writings. But his claim that the creation accounts had "picture language" rocked me. In fact, after the lecture, about fifty of the seventy or so students stormed to the front of the class to confront this professor.

We were deeply troubled by the idea of "picture language" in the Genesis creation accounts. It challenged a major assumption we made about Scripture. Like most Bible-believing Christians, we assumed that God had revealed some basic scientific facts in Scripture well before their discovery by modern science. This assumption is known as "concordism," or more precisely "scientific concordism." The word "concord" means "to be in agreement" or "in harmony." Scientific concordism is the belief that there is an alignment between the Bible and the facts of science.

In assuming that God revealed scientific information in Scripture, most of us in the class believed this was proof that the Bible really is the Word of God. Only a God who is powerful and transcends time could have given modern scientific facts to the ancient authors of Scripture. It is easy to understand why we reacted negatively to our professor's claim that Genesis had "picture language." We interpreted this as an attack on the Word of God and, ultimately, as an attack against God.

Scientific concordism is a reasonable assumption. After all, God is the Creator of the world and he is also the Author of the Bible. Therefore, to expect some sort of alignment or harmony between the Book of God's Works and the Book of God's Words makes perfect sense. It is worth noting that most Christians today are concordists. But my theology professor was raising some disturbing questions in my mind. Does the Bible really align with the scientific facts found in nature? More importantly, if Scripture and science don't match up, will this destroy my belief that the Bible is the inspired Word of God?

Challenges to scientific concordism continued during the next term in a course on the relationship between science and Christianity. Most of the students were six day creationists. We battled our "liberal" professor and his guest lecturers because they believed the earth was extremely old. Frustrated by the one-sidedness of the course, I invited and paid for a well-known creation scientist to come and lecture so we could have some balance in what we were being taught.

Then one day after class I cornered my professor in a hallway and asked him directly, "What do you think about the idea that the world was created in six literal days about six thousand years ago?" He answered bluntly, "It is an error." I can still remember how the word "error" rattled my soul. I knew this professor was a remarkable Christian because I had taken a course from him in the previous term. I had great respect for his honesty. But he did not believe in six day creation! This was the very first time in my life I had met a *real* Christian who said that creation in six days is wrong.

On the final day of the course, the professor looked directly at me and asked, "Denis, if you gave up your belief in six day creation, would you also give up your faith in Jesus?" Wow! That was one question I was not expecting. Looking back at that moment, I am convinced that it wasn't my professor talking to me. The Lord was speaking through him to challenge my understanding of the Christian faith. In attempting to respond, I mumbled and stumbled and never really answered. Deep in my heart of hearts, I knew that my personal relationship with Jesus was so much more important than any view on how God had created the universe and life.

I stepped away from this science and Christianity class still believing in six day creation. But for the first time, I began asking myself whether a literal reading of Genesis 1 is an error. I also wondered about the truthfulness of scientific concordism. Is it possible for a Christian to reject the assumption that God has revealed basic scientific facts in the Bible? I wasn't sure.

Genesis 1: Ancient Science and Ancient Poetry

Theology school made me rethink how God had inspired the writers of the Bible. As I continued to study Genesis 1–11, it became evident that scientific concordism is not a feature of Scripture. Instead, I started to discover evidence *within the Bible itself* that there was an ancient understanding of the physical world. Or to say it another way, the Book of God's Words includes what could be termed "ancient science." Let me give you an example.

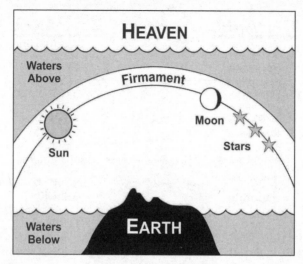

Figure 2–1. Ancient Understanding of the Structure of the Universe.

Figure 2–1 illustrates the structure of the universe found in Scripture. This ancient science appears in Genesis 1 with the creation of the heavens. On the second day of creation, "God said, 'Let there be a firmament between the waters to separate waters from waters.' So God made the firmament, and divided the waters under the firmament from the waters above it. And it was so. God called the firmament 'heaven'" (Gen. 1:6–8).[1] The original biblical word translated as "firmament" means "a hard dome."[2] Why did ancient people believe this? Well, think about it. When they looked up at the sky, they saw a massive blue dome. To conclude there was a heavenly body of water held up by a solid structure made perfect sense.

On the fourth day of creation, "God said, 'Let there be lights in the firmament of the heaven to divide the day from the night; and let them serve as signs to mark sacred times, and days and years; and let them be lights in the firmament of the heaven to give light on the earth.' And it was so. God made two great lights—the greater light to govern the day and the lesser light to govern the night. He also made

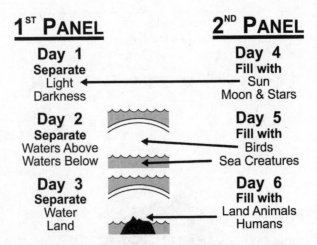

Figure 2–2. Parallel Panels in Genesis 1 Creation Account.

the stars. God set them in the firmament of the heaven to give light on the earth" (Gen. 1:14–17). Again, from the point of view of the naked eye, this is what it looks like. The sun, moon, and stars are right up against the blue dome, giving the appearance that they are embedded in its surface. In fact, this was the best science-of-the-day in the ancient world.[3] (In chapter 5 we will examine more examples of the ancient science in Scripture.)

During my theological studies, I also discovered that there is ancient poetry in the Bible. It's important to define the term "poetry." It simply means "structured writing." For example, the creation account in Genesis 1 is built on an ancient poetic structure called "parallel panels" as shown in Figure 2–2.

In the first panel, God forms the boundaries of the universe. He then fills the world with heavenly bodies and living creatures in the second panel. The panels are parallel to each other. On the first day of creation, God creates light in alignment with the sun, moon, and stars being placed in the firmament on the fourth day. The separation of the waters above from the waters below on day two creates an air space for birds and a body of water for sea creatures, both made on day five.

Finally, on the third day God commands dry land to appear, matching up with the creation of land animals and humans on the sixth day.

Discovering ancient science and ancient poetry in Scripture was quite troubling for me at first. Yet it became clear that Genesis 1 could not be a literal and scientific account of how God actually made the world. We do not live in a universe with a firmament that holds up a heavenly body of water above us. The sun, moon, and stars are not embedded in a hard dome. And real events in the world do not happen following parallel panels. Slowly I came to the conclusion that scientific concordism is not a feature of the Word of God. Stated another way, the Bible is not a book of science.

Viewing Scripture in a New Light

Studying theology introduced me to a new way of thinking about how God had inspired the biblical authors. This is summarized in one famous saying that I learned. *The Bible is the Word of God written in the words of men and women during ancient times.*[4] To be sure, Scripture definitely contains "the very words of God," as the apostle Paul states in Romans 3:2. Yet, for the Lord to reveal himself to ancient people in the past, he came down to their intellectual level. In doing so, God used their understanding of nature (ancient science) and their writing techniques (ancient poetry) as vessels to deliver life-changing spiritual truths.

The idea that God descends to our level to communicate with us is known as the "Principle of Accommodation." The word "accommodate" means "to adapt," "help out," and "make suitable." This happens naturally when we speak with small children. We go down to their level of understanding and use simple words and simple ideas so they can grasp what we are saying.

Just think about it for a moment. God is the infinite Creator, and we are finite creatures. For us to understand him and for him to communicate with us, the Lord has to descend to our level and use ideas we can comprehend. I am sure every Christian will agree that the mind of God is so much greater than the mind of humans. As God states in Isaiah 55:8–9, "For my thoughts are not your thoughts. . . . As the heavens are higher than the earth, so are my ways higher than your ways and my thoughts than your thoughts." I am also certain that all Christians experience God's act of accommodation when he speaks to us in prayer. Doesn't the Lord meet you exactly where you are in your life, and doesn't he talk to you at your level?

This is also the case with the Bible. It was written in ancient times by an ancient people who did not have modern scientific instruments like telescopes or microscopes. They had no idea that the universe is incredibly huge or that it contains billions of galaxies. These men and women did not know how to date the age of rocks, and they were not aware of all the fossils in the rock layers of the earth. There was no reason for them to think the world was old or that living organisms had evolved. Of course, God could have revealed in Scripture that he created the world through evolution over billions of years. But I doubt that ancient people would have grasped such a concept.

Instead, God was accommodating in the creation accounts of the Bible. He allowed the inspired writers to use their ancient scientific ideas about origins to reveal the foundational message of faith that he alone was the Creator of the entire world (Gen. 1:1). This ancient science-of-the-day is a vehicle that also transports the spiritual truths that the creation is "very good" (v. 31) and that only humans were created "in the image of God" (v. 27). In my opinion, this is another amazing example of how much God loves and cares for us. He came down from heaven to reveal himself by using ancient human ideas about nature and origins so that we could understand his spiritual messages for our life.

And isn't this exactly what Jesus did? God lowered himself and became a man in the person of Jesus to show us that he loves us more than we could ever imagine. The Lord also accommodated in his teaching ministry by employing parables. These are earthly stories that every ancient person would have understood. Jesus used them as vessels to reveal heavenly truths.

Studying to become a theologian was one of the most challenging experiences of my life. Shockingly, it was the evidence *within the Bible itself* that crushed my dream of becoming a creation scientist. Even though I came to reject six day creation, my faith in God and the Bible were as strong as ever. My love for Jesus and Genesis 1 were the very same as when I was a six day creationist. Praise the Lord for his sustaining grace! I realized that my personal relationship with God was not based on how he had made the universe and life. Rather, it was founded on his love for us and my experience of walking with Jesus every day.

Opening the Book of God's Works

I realized that the Bible is not a book of science, but my original "Grand Plan" to destroy the theory of evolution was still alive and well. I moved on to obtain a PhD in biology. Since I was a dentist, I entered a university program to study the so-called "best evidence" for evolution—the evolution of teeth and jaws. My plan was to collect scientific facts that disproved evolution, and once I graduated as a scientist, I would write a devastating book against Satan's lie that life had evolved.

Well, you can probably figure out what happened. My soul was shaken to the core for a second time. I began to see fossil evidence that indicated evolution was true. For years in Sunday school and at scientific creationist events, I had been taught that there were no transitional fossils. These are fossils of extinct animals that appear in between two different kinds of extinct animals in the rock layers of the earth. Transitional fossils share characteristics with the animals from which they evolved and also have features of the animals into which they evolve. Therefore, these fossils are evidence of the transformation of one type of creature into another creature.

During my scientific training, I saw and even held in my hands a number of transitional fossils. When I first discovered that these fossils existed, it was not at all comfortable. I tried my best to explain their existence through an anti-evolutionary view of origins. But I could not deny this scientific evidence in the Book of God's Works, and eventually I accepted evolution.

Early Amphibian 360 mya

Lobe-Finned Fish 385 mya

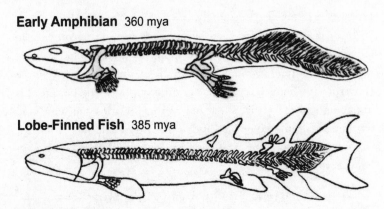

Figure 2–3. Fish-to-Amphibian Evolution (mya: million years ago).

In his famous book *Evolution: The Fossils Say No!*, Duane Gish claimed that "the discovery of only five or six of the transitional forms scattered through time would be sufficient to document evolution."[5] Let's look at some examples of transitional fossils that led me to believe that living organisms had evolved. You can decide whether there are "five or six transitional forms" that meet Gish's requirement for proving that biological evolution is true.

Fish-to-Amphibian Evolution

The theory of evolution states that fish evolved into amphibians. Figure 2–3 presents an ancient lobe-finned fish and one of the first amphibians.[6] A lobe fin is a thick fleshy fin with bones inside it. As you can see, this fish and amphibian have similar heads, back bones, and tails. In fact, the early amphibian looks like a fish with legs.

Remarkable fossil evidence for the evolution of fish into amphibians comes from the transition of fins into limbs as seen in Figure 2–4.[7] In contrast to modern fish such as bass and trout that have only thin strips of bone in their fins, the fins of lobe-finned fish had large limb-like bones. These fins gave fish the advantage of greater mobility that eventually allowed them to come onto land and find new sources of food.

As the fin in the lobe-finned fish evolved, finger-like projections

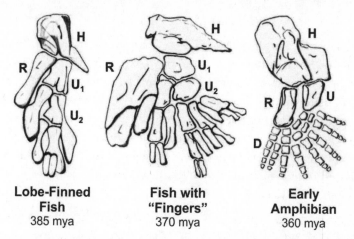

Lobe-Finned Fish
385 mya

Fish with "Fingers"
370 mya

Early Amphibian
360 mya

Figure 2–4. Fin-to-Limb Evolution. (H) humerus (R) radius (U) ulna (D) digits (mya: million years ago).

appeared in a fish named the "fish with 'fingers.'" You will note that there are 8–9 finger-like bones in this fish and that the early amphibian limb has 8 fingers. That's no coincidence. It is scientific evidence that the first amphibians evolved from fish with a similar number of finger-like bones in their fins.

Another feature I found fascinating is a similarity in the teeth of lobe-finned fishes and early amphibians. As you know, teeth are made up of an outside layer of enamel that covers a core of dentine. Inside the dentine there is a pulp canal containing nerves and blood vessels. In nearly all toothed animals, this canal is smooth and tubular. However, lobe-finned fishes and the first amphibians have a maze-like pulp canal as shown in Figure 2–5.[8] This is a

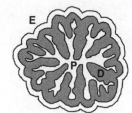

Figure 2–5. Tooth with Maze-Like Pulp Canal.
(E) enamel (D) dentine (P) pulp canal.

very rare dental feature and indicates that as fish evolved into amphibians, these unique teeth in lobe-finned fishes were passed on to the earliest amphibians.

Reptile-to-Mammal Evolution

Evolutionary scientists have discovered that mammals evolved from reptiles. Some of the best fossil evidence comes from the transition of ancient reptile teeth into early mammal teeth presented in Figure 2–6.[9] Most reptiles have simple cone-shaped, single-rooted teeth that are all about the same size. These teeth function well for grasping and killing animals, but they are not useful for chewing. Consequently, reptiles swallow their prey whole or in large chunks and do not draw all the nutrients from the flesh of victims. That's why our moms told us to eat slowly and chew our food!

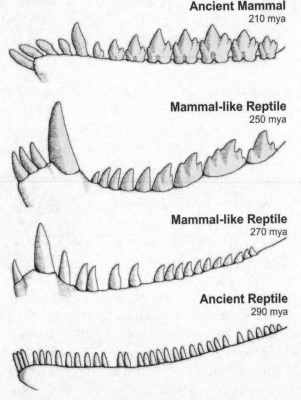

Ancient Mammal
210 mya

Mammal-like Reptile
250 mya

Mammal-like Reptile
270 mya

Ancient Reptile
290 mya

Figure 2–6. Reptile-to-Mammal Tooth Evolution (mya: million years ago).

As reptiles evolved into animals known as "mammal-like reptiles," a special tooth began to lengthen at the corners of the mouth. This created a weapon that was useful for stabbing prey and tearing up their flesh. You are correct to suspect that this tooth eventually became the prominent canine tooth we see in dogs and lions today. As dental evolution continued, the back teeth became wider and sharp points termed "cusps" started to appear. This created a set of teeth that could cut up flesh and get more nourishment out of it.

Early mammals then appeared with very specialized back teeth. Those that were farthest back in the mouth had two roots. This produced teeth that were much more stable and allowed for the grinding of food and drawing out of more nutrition. In addition, the sharp cusps on the back teeth of the upper and lower jaws lined up across from each other and interlocked tightly together. This created a dentition that was more effective in slicing up prey, and early mammals were able to extract even more vital nutrients.

Further evidence for the evolution of reptiles into mammals appears with the transition of their jaws as seen in Figure 2–7.[10] Reptiles have a lower jaw made up of a number of bones. Their jaw joint connects the quadrate bone at the base of the skull with the articular bone of the lower jaw. However, in mammals like us, the lower jaw is only one bone and the jaw joint is between two entirely different bones—the squamosal bone and the dentary bone. I'm sure you are asking the question: how could reptiles have evolved into mammals with entirely different jaw joints when an animal needs a functioning jaw joint to eat and live?

The answer is found in the fossil record of mammal-like reptiles. As the jaw of reptiles evolved, the dentary bone increased in size and the other bones got smaller. Then amazingly, some mammal-like reptiles developed another jaw joint, giving them two jaw joints! One was the original joint of the reptile and the other was a new and small mammalian joint between the dentary and squamosal bones. This is an excellent example of a transitional fossil, since it features distinct characteristics from both reptiles and mammals. Over time the quadrate and articular

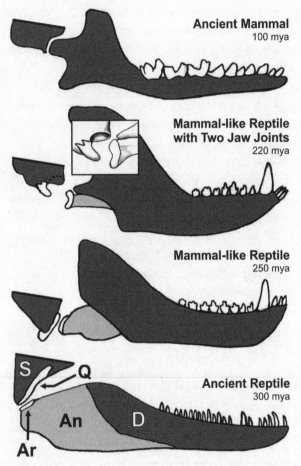

Figure 2–7. Reptile-to-Mammal Jaw Evolution. (An) angular (Ar) articular (D) dentary (Q) quadrate (S) squamosal (mya: million years ago).

bones separated from the jaws, eliminating the reptilian jaw joint. The dentary bone then became the lower jaw in mammals.

It is worth pointing out that the fossil record of the evolution of reptiles into mammals is quite complete. It is so gradual in places that it is difficult at times to classify different fossils. The fact that the term "mammal-like reptiles" was coined indicates these animals have both reptilian and mammalian features. In other words, mammal-like reptiles are scientific proof that transitional fossils really do exist.

Land Mammal-to-Whale Evolution

Here are a few more examples of transitional fossils and evidence for evolution. The theory of evolution asserts that whales evolved from land mammals that entered the oceans. It is important to remember that whales are not fish. They are mammals like us. Whales have hair, are warm-blooded, and develop in the womb of their mother. The females also have breast glands to nourish their young with milk.

Figure 2–8 presents four examples of transitional fossils in the evolution of ancient whales.[11] Although the earliest whales had short limbs, they were still able to walk on land. Their feet were huge, and

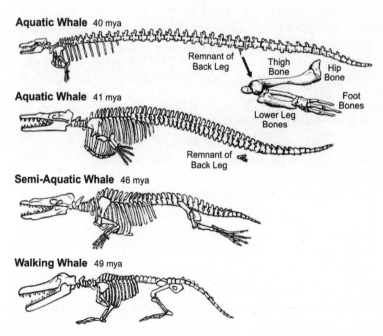

Figure 2–8. Ancient Whale Evolution (mya: million years ago).

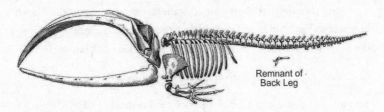

Remnant of
Back Leg

Figure 2–9. Modern Toothless Whale.

this improved their ability to swim. Notably, the walking whale had hooves on its feet just like cows! This feature is evidence that whales evolved from land mammals. Over time the bodies of ancient whales got more streamlined, and their tail lengthened to become the main power source for thrusting themselves through water.

As ancient whales evolved, their back legs were reduced until they became small remnants of bone. These non-functional malformed legs continue to appear in modern whales as shown in Figure 2–9.[12] Body structures like remnant hind legs are known as "evolutionary vestiges." The word "vestige" means a "trace" or a "leftover." Thus the tiny deformed legs in modern whales are a leftover trace of evidence indicating that whales evolved from land mammals with back legs.

Another fascinating evolutionary vestige appears with whale teeth. As Figure 2–8 reveals, ancient whales had teeth because they evolved from toothed land mammals. But some whales today do not have teeth, like the whale in Figure 2–9. Instead, these toothless whales have a massive mouth that filters huge amounts of water to collect small sea creatures. Here's the interesting thing: when a toothless whale develops in the womb, teeth are made. But as Figure 2–10 illustrates, these embryonic teeth are malformed and never develop complete roots.[13] In fact, they do not attach to the jaws and are lost before birth or soon after. Being mammals, these developing whales are nourished through the umbilical cord in the womb. Once they are born, they drink milk from their mother's breast glands. Therefore, these baby whales have no need for teeth.

Here are a couple questions I would like you to think about.

Figure 2–10. Embryonic Teeth of Toothless Whales.

If God created the great sea creatures like whales on the fifth day of creation in Genesis 1, why did he place useless little back legs in numerous species of whales? Or why did the Creator put the genes to make teeth in toothless whales? For me, the answer is simple. These miniature legs and tooth genes are evolutionary vestiges. They are scientific evidence that evolution is true.

It is important to emphasize that I've presented only a handful of transitional fossils. The fossil record actually has an incalculable number of transitional creatures that demonstrate living organisms evolved over time. You can learn about them in any good science library.

I also want you to know that every time a new transitional fossil is discovered, it *always* appears in the rock layer of the earth where the theory of evolution predicts it should be. For example, toothless whales are never found in lower layers before the appearance of the first toothed whales. Think about that. Every year there are thousands of new fossils found, and they *always* show up exactly where scientists expect them to be in the fossil record. This too was powerful evidence that convinced me that evolution is true.

The Embryology-Evolution Analogy

Alongside my research on evolution, I studied embryology. This science examines the formation and development of living organisms. It was amazing to learn about the incredibly complex set of well-coordinated natural processes involved during the early stages of life in different animals. In particular, I worked on the dental development of tadpoles. Yes, they have teeth, but only in the upper jaw. What is remarkable about these creatures is that their skin is transparent. After putting a tadpole to sleep with an anesthetic, I could chart the growth of the teeth with a microscope. I was also able to observe its organs, muscles, and nerves, and I even saw red blood cells being pumped out of the heart into the arteries!

My study of embryology filled my soul with awe and strengthened my faith in a Creator. This scientific evidence made it obvious to me that living organisms are magnificently designed and point to an Intelligent Designer. Psalm 19:1 states, "The heavens declare the glory of God." I would add to this verse that humble little tadpoles also "proclaim the work of his hands"!

In my research of both embryology and evolution, the Lord introduced me to a significant similarity between these two sciences, known as the "Embryology-Evolution Analogy." I began to see a parallel between God's creative action in the formation of every individual living organism today, and his creative activity during the evolution of all living creatures in the past. Let me explain.

As a Christian, I believe that God creates each of us by using his natural process of embryology. As Psalm 139:13–14 asserts, "For you [God] created my inmost being; you knit me together in my mother's womb. I praise you because I am fearfully and wonderfully made." I have yet to meet a Christian who thinks that the Creator came out of heaven to miraculously attach an entire leg or arm to their developing body in the womb. Instead, we believe that the Lord creates every person through his *ordained* and *sustained* embryological mechanisms. In this way, these laws of nature have been ordered by our Maker, and he has upheld them each moment during our development.

In light of our belief that God made us by employing natural processes in the womb, here are some questions I would like you to consider: Is it possible that the Lord created another set of creative mechanisms in nature that scientists call "evolutionary processes"? Could it be that instead of coming out of heaven and miraculously placing each creature on earth, God "knit together" all living organisms through his ordained and sustained natural process of evolution?

And if this is the case, can we say that evolution is an intelligently designed mechanism, since it led to the creation of plants and animals that are "fearfully and wonderfully made"?

Discovering the similarity between God's creative action in embryology and evolution completely *freed* me from being afraid of evolution. It became evident that science is the study of the Lord's creation and all the natural mechanisms that he created, including the process of evolution. Instead of being an enemy of Christianity, science is a gift from our Creator that declares his glory and reveals to us how he made the universe and life.

By opening the Book of God's Works to the chapters on evolution, I came to the same conclusion held by nearly everyone who has actually studied evolutionary biology: the evidence for evolution is *overwhelming*. Twenty years ago I would never have believed that one day I would say this. But we need to follow the scientific evidence no matter where it leads. We must thank God for giving us the ability to practice science and understand his creation. And I thank the Lord for the privilege of studying biological evolution and discovering how he created living organisms.

Final Thoughts

That's my story. No doubt about it, I struggled for a long time trying to make sense of the relationship between evolution and Christianity. As a born-again Christian, I can say that my views on origins changed a lot over the years, but my belief in God never changed. The Jesus I loved and served when I accepted six day creation is the very same Jesus I love and serve today as someone who believes the Lord created the world through evolution. As Hebrews 13:8 states, "Jesus Christ is the same yesterday and today and forever." Amen!

My love for the Bible hasn't changed either. Every morning I begin my day by reading the Word of God. Like a spring of pure water, Scripture quenches the spiritual thirst of my soul (John 4:10). It also remodels and transforms my mind in order for me to discover God's will for my life (Rom. 12:2). If anything has changed, my training in theology has helped me to focus more on the marvelous life-changing spiritual truths in the Bible. As Isaiah 40:8 reveals, "The word of our God endures forever." Double amen!

There is a question that I suspect arose in your mind while reading my story. When I was in medical school, I believed that God called me away to become a creation scientist to attack evolutionists in universities. What am I to make of this calling, because after opening God's Two Books, it became obvious to me that the Creator did not make the world in six literal days six thousand years ago?

Today I still believe that I was called by the Lord at that time. But in doing so, he accommodated and came down to my spiritual and intellectual level by using the only Christian view of origins that I understood—six day creation. As many of you have experienced, God meets us exactly where we happen to be in our life. When he called me, I was trapped in "either/or" thinking and was stuck in the origins dichotomy. The Lord had a plan for my life, and it included an education in his Two Books. In retrospect, I now see that God did indeed call me to attack atheistic interpretations of evolution and defend the belief that the world is his creation.

To conclude, I hope that you will agree with me that the origins dichotomy is a mistake. Stated more precisely, it's a *false dichotomy*. It limits our choices and forces us to believe that there are only two credible positions on origins. As I suggest throughout this book, there are ways to free us from "either/or" thinking and to get beyond the so-called "evolution vs. creation" debate. To start this process, we need to define some terms.

CHAPTER 3

TERMS THAT BEGIN
TO FREE US

Words are powerful. I can remember the day many years ago when my high school English teacher made that point. Intuitively, I knew I was being taught something that was quite significant. As I shared my story of struggling with origins in the previous two chapters, you might have asked yourself what exactly do the words "creation" and "evolution" mean?

For the longest time, I believed that the term "creation" referred only to the world being made in six 24-hour days about six thousand years ago. I also assumed that "evolution" had to be connected to atheism. Do you see the problem and the power of the common everyday meaning of these terms? They forced me into thinking that there were only two positions on origins. I had to choose *either* evolution *or* creation.

The origins dichotomy is fueled by another problem—the conflation of ideas. The term "conflation" is derived from the Latin prefix *con-*, which means "together," and *flare*, the verb "to blow." Think of two people blowing into a balloon. The air molecules from each person mix together completely. Conflation refers to the careless blending of distinct concepts into one single, poorly defined idea.

For most people today, the word "evolution" is conflated with a godless and purposeless view of the world. It forces them to believe it is impossible for an evolutionist to believe in God and, in particular, the God of Christianity. Similarly, the term "creation" is conflated with the literal interpretation of Genesis 1. This notorious conflation drives many Christians into assuming that they must reject evolution and accept six day creationism.

The problem with conflation is that it narrows the meaning of words and limits the range of possibilities regarding an issue such as origins. My story of moving back and forth between so-called "evolution" and "creation" demonstrates that my thinking was controlled by conflated misunderstandings of these words. As a corrective, this chapter introduces terms that free us from common conflations and assist us in getting beyond the "either/or" thinking of the origins dichotomy. You will then begin to outline your position on origins.

Creation

A majority of people today conflate the word "creation" with a strict literal reading of Genesis 1 in which God made the world in only six days. It is often assumed that this is *the* official view of origins in Christianity. Some will go so far as to say that to be a *real* Christian, we must reject evolution and believe in six day creation. As I revealed in the first chapter, there was a time when I believed this was true.

Let me suggest that to define the term "creation," we should consider the way theologians use it in their day-to-day work. They emphasize that this word is a religious idea and not a scientific concept. More precisely, it is a religious *belief* and refers only to the things God made. Someone who is a "creationist" is a person who simply believes in a Creator and that the entire world is his creation. Theologians are interested in how the world originated, but they know that science is not their area of expertise. Therefore, it is important for them to have trusted Christian friends who are scientists to help them understand the Lord's creative method.

The work of theologians has led to the formulation of the doctrine of creation. This doctrine is based on the Bible and includes seven basic religious beliefs about the creation and the Creator.

- The creation is completely separate from the Creator (Gen. 1:1; John 1:1–3). The world is not God, and no part of it is divine. The Creator is above and beyond the creation. In other words, God transcends the world he made.

- The creation is totally dependent on the Creator (Acts 17:24–28; Rev. 4:11). God *ordained* (ordered) the world into existence and continues to *sustain* (uphold) it every single instant today (Ps. 65:9; Heb. 1:3). The Creator, then, is also immanent and ever-present within his creation.

- The creation was made out of nothing (Col. 1:16–17; Heb. 11:3). Only God existed before anything was created and brought into existence.

- The creation has a beginning (Gen. 1:1; Heb. 1:10) and an end (Heb. 1:11–12; 2 Peter 3:10–13). The Creator not only made the physical world but also created time. Therefore, the present universe has not always existed, nor is it going to last forever. Only God is eternal.

- The creation declares the existence of the Creator (Ps. 19:1–4; Rom. 1:19–20). The world's beauty, complexity, and functionality reflect intelligent design and point to an Intelligent Designer.

- The creation is very good (Gen. 1:31). The world is the ideal setting that allows men and women to develop a personal relationship with the Creator. Since the creation is very good, we need to take care of it.

- The creation features humans who are created in the Image of God (Gen. 1:26–27). We are God's greatest creation. Men and women are unique, and no other creature enjoys such an honored status, because we have been made in the likeness of our Creator.

To conclude, the term "creation" is a religious belief. For theologians, this word simply refers to everything the Creator has made and not to the method he used to create the universe and life. Stated in another way, the Christian doctrine of creation does not deal with *how* the world was created, but rather focuses on *who* created it.

--- Evolution ---

For most people today, the word "evolution" is conflated with the notion that the world arose through a natural process driven only by blind chance. From this perspective, there is no place for God, and our existence has no ultimate meaning or purpose. Humans are merely a fluke of nature. Regrettably, the conflation of evolution with atheism is often assumed to be *the* official view of modern science. Some even go so far as to claim that *real* scientists have to be atheists. It's not surprising that many believe it is impossible to be both an evolutionist and a Christian. This is another misguided idea I once embraced many years ago.

But let's define the term "evolution" in the way most scientists use it in their day-to-day work in the laboratory. Evolution is a scientific idea, or more precisely, a scientific theory. It simply asserts that the cosmos and living organisms, including humans, arose through natural processes over billions of years. There is no mention of whether these processes in nature were created by God. The reason for this is because science deals only with physical reality and not spiritual reality.

To further explain the notion that science is limited to investigations of the natural world, think about the force of gravity on earth. Most of you know the scientific formula.

$$F = mg$$

"F" is the force of gravity that pulls objects toward the earth; "m" is the mass of the object; and "g" is the acceleration of the mass downward due to gravity. There is no mention in this formula of God or God's activity in gravity. Why? Science is restricted to the physical and does not deal with the spiritual.

Now it is important for me to make something extremely clear. I believe that God created gravity. It is one of his ordained and sustained natural processes in the world. If we didn't have gravity, everything, including us, would be thrown off the earth as it spins on its axis. Therefore, gravity is a very good creation. In science, there are no scientific instruments to detect God and his activity in nature. Again, it's because science deals only with the physical world and not the spiritual realm.

To complete the definition of the term "evolution," we need to identify three basic evolutionary sciences.

- *Cosmological Evolution* examines the origin of stars, planets, and moons. Cosmologists have discovered that the universe began with a massive explosion about 13.8 billion years ago. Termed the "Big Bang," this event led to the origin of space, time, and matter and eventually to the evolution of all the astronomical bodies in the cosmos.

- *Geological Evolution* investigates the formation of the earth. Geologists have demonstrated that our planet is around 4.5 billion years old. They study natural processes that are going on in the earth today, such as erosion, volcanic activity, and the movement of continents. From this scientific evidence, geologists are able to describe how the earth has evolved and changed over time.

- *Biological Evolution* explains the origin of living organisms. Fossils found in the rock layers of the earth reveal a pattern that shows how plants and animals gradually transformed over time into entirely new species. Biologists explore the natural processes that organized simple molecules into the first forms of life about 3.8 billion years ago and that later evolved into every creature that has lived on earth. To illustrate by using the vertebrates (animals with a backbone), evolutionary biologists have shown that fish evolved into amphibians, which changed into reptiles, and these evolved into mammals, which include humans, who appeared last in evolution.

Even though these three sciences dealing with evolution do not make any mention of God, I firmly believe that every natural process discovered by evolutionary scientists was *ordained* by the Creator. I also believe that God *sustained* these processes during the billions of years of cosmological, geological, and biological evolution.

Let me offer an analogy to explain how I view the creation of the world. Imagine God's creative action is like the stroke of a pool cue in a game of billiards. Divide and label the balls into three groups using the words "heavens," "earth," and "living organisms," and let the eight ball represent humans. In depicting the origin of the world, a six day creationist sees the Creator making single shot after single shot with no miscues until all the balls are off the table. No doubt about it, that's impressive.

However, as a Christian evolutionist, I picture God using only one stroke of his cue representing the Big Bang. His opening shot is so incredibly precise that not only are all the balls sunk but they drop in order. The balls labeled "heavens" fall first, then "earth," followed by "living organisms," and finally the eight ball—the most important ball in billiards—signifying human beings. To complete the analogy, the Lord pulls this last ball out of the pocket and holds it to his heart to indicate his personal relationship with men and women.

Isn't the Creator who uses just a single stroke to sink all the balls infinitely more amazing than the God of the six day creationists who takes shot, after shot, after shot? I believe that the Lord's eternal power and unfathomable foresight is best illustrated by creating through an evolutionary process that he set in motion with the single miraculous event of the Big Bang. Just think about it. God with only one creative act set up the laws of nature for everything in the entire world to self-assemble through evolution. I can't think of a greater example of intelligent design.

To conclude, the term "evolution" refers to a scientific theory. For scientists, evolution simply describes the origin of the world over billions of years through natural processes. Therefore, evolution explains *how* the universe and life originated. It does not reveal *who* began the evolutionary process.

Teleology and Dysteleology

The words "teleology" and "dysteleology" are terms most people do not know. However, I am convinced that nearly everyone is aware of the basic ideas behind these words. In my opinion, teleology and dysteleology are two of the most important concepts we need to master to understand origins.

The term "teleology" comes from the Greek noun *telos* and it has a number of meanings, like "plan," "purpose," "end," and "final goal." A person who believes that the world features teleology is someone who believes there is an ultimate plan or purpose for our existence and that we are moving toward an end and final goal. For example, I am a teleologist, and I base my teleology in the God of Christianity. The Bible reveals that the universe has an overall plan and purpose that is rooted in the will of our Creator. I also believe the world is moving toward an ultimate goal in which those who love God will be with him throughout eternity.

In contrast, the word "dysteleology" refers to the belief that the world does not have any ultimate end or purpose. This bleak and dismal picture claims our existence is ultimately pointless and that there is no ultimate right or wrong. The most famous atheist today is Richard Dawkins. He is the best example of a dysteleologist. With brimming confidence he proclaims, "The universe we observe has precisely the properties we should expect if there is, at bottom, no design, no purpose, no evil and no good, nothing but blind, pitiless indifference."[1]

If you will recall my personal story in chapter 1, I became an atheist by the time I finished dental school. Even though I was not aware of the term "dysteleology," I was definitely a committed dysteleologist. I did not believe in the existence of God, love, or morality, or that the world had any ultimate purpose or final end.

Now equipped with the terms "creation," "evolution," "teleology," and "dysteleology," let's identify some relationships between them as shown in Figure 3–1. Most people today are indoctrinated by both the church and secular society into believing that evolution is by necessity dysteleological. They assume evolution has to be an unplanned and purposeless natural process driven only by blind chance. Of course, this is a notorious conflation. The scientific

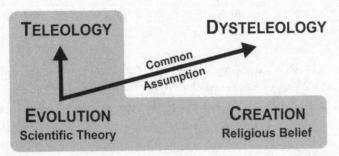

Figure 3–1. Terms and Relationships. The shaded area introduces the notion of teleological evolution and the possibility of accepting the scientific theory of evolution and the religious belief in a Creator, with the world being his creation.

theory of evolution is carelessly blended with a belief in dysteleology. Therefore, we must qualify this interpretation of evolution and call it "dysteleological evolution."

Let's challenge this common assumption that evolution is necessarily dysteleological. Why couldn't evolution be teleological? From this point of view, evolution is a planned natural process that heads toward a final goal—the creation of the universe and life with men and women. If this is the case, the Creator made the world through "teleological evolution."

From this perspective, God's creative action was not through dramatic and miraculous interventions forming each individual star, planet, and moon, or every plant and animal. Instead, the Lord created the world by using an ordained and sustained evolutionary process. And should you hold this position, you are *both* a creationist by believing in a Creator and an evolutionist in accepting the natural process of evolution. As we will see in chapter 6, Christians who embrace this view of origins are known as "evolutionary creationists."

Intelligent Design

The term "intelligent design" (ID) is often heard in our churches today. This is due to the powerful influence of a group of Christians promoting what is called "Intelligent Design Theory." Leaders of this view of origins include Phillip Johnson, Michael Behe, William Dembski, and Stephen Meyer.[2] Their central claim is that design in nature is *scientifically detectable*. ID theorists reject biological evolution. They argue that God used miraculous interventions to place design in living organisms.

But can you see the problem the leaders of ID Theory have created for us? These Christians have produced another dichotomy: biological evolution vs. intelligent design. They are forcing us into "either/or" type of thinking and making us pick between evolution and design. But if God created the world through teleological evolution, intelligent design could be a characteristic of the evolutionary process.

In the last chapter, I introduced the Embryology-Evolution Analogy (see page 41). During my training in science I was astonished to discover the complex set of natural processes that create each living organism. These embryological mechanisms inspired me to believe that they were designed by a Creator. This could also be the case with evolution. Think about it. Beginning with the Big Bang, God put in motion extremely well-designed natural processes which he used to self-assemble the entire world, including us. Only a Creator with unbelievable power and incredible foresight could have designed such an evolutionary process.

Let me now offer the traditional definition of the term "intelligent design." Intelligent design is a *belief* that the world's beauty, complexity, and functionality point toward an Intelligent Designer. Have you ever been out in nature and been overwhelmed by how beautiful it is? I am sure you have. God is like a Cosmic Artist who has painted a breathtaking panorama on the canvas of creation. Have you ever studied the different cells in the human body? They are incredibly complex with countless well-coordinated parts and processes that function automatically every second. It seems clear to me that only a Supreme Engineer could have constructed all the mind-boggling cells in our body.

The Bible also affirms the reality of intelligent design. Let me briefly cite two of the most important passages that claim nature points to God, and we will examine these in more detail in the next chapter. Psalm 19:1 states, "The heavens declare the glory of God; the firmament proclaims the work of his hands."[3] In other words, our spectacular world shouts out to us that it was made by a Creator! I find this to be so true. When I look at the natural world, it always strengthens my belief in an Intelligent Designer, who for me is the God of Christianity.

Similarly, Romans 1:20 asserts, "Since the creation of the world God's invisible qualities—his eternal power and divine nature—have been clearly seen, being understood from what has been made, so that men and women are without excuse."[4] The creation not only points toward God, it even reveals to us some of his amazing attributes. He is incredibly powerful, and he is a divine being. Romans 1:20 states that the message inscribed in the creation is so obvious that everyone can understand it. There's no excuse for not believing in the Creator.

To summarize, intelligent design is a religious belief. It is not a scientific theory. And no, design is not scientifically detectable. There is no scientific instrument that can detect intelligent design in nature, like a Geiger counter that detects radiation. To state it again, science deals with the physical, not the spiritual. However, design in nature is a reality. God has made our minds in such a way that when we look at our spectacular world, it impacts us powerfully and points us toward him.

The Metaphysics-Physics Principle

In light of the terms that we have defined above, I can now propose a concept that is foundational for developing a fruitful relationship between modern science and Christian faith—the Metaphysics–Physics Principle as outlined in Figure 3–2. To move beyond dichotomies and conflations, it is necessary to respect the fundamental differences between scientific discoveries and our ultimate religious and philosophical beliefs. In this way, we can begin to view science and Christianity in a complementary relationship.

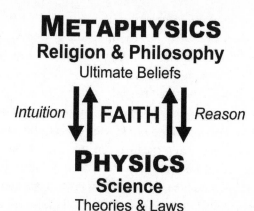

Figure 3–2. The Metaphysics-Physics Principle.

The word "complementary" comes from the Latin verb *complēre*, which has the meanings "to fill," "complete," and "perfect." In a complementary relationship between two parties, each adds something that is lacking in the other so that they enrich and enhance each other. From my perspective, Christian faith and modern science fulfill and strengthen each other. They lead us to a completely integrated understanding of the Creator, his creation, and our existence. In the final chapter, I will come to the conclusion that there is a complementary relationship between the Book of God's Words and the Book of God's Works.

As we have noted, science deals only with physical reality. Through observations and experiments, scientists formulate theories and laws about the structure, operation, and origin of the natural world. To employ the Greek word for "nature," science investigates *phusis*. The English terms "physics" and "physical" are derived from it.

When we think about all the incredible scientific discoveries today, we cannot help but ask questions about their ultimate meaning. In other words, everyone wonders about the religious and philosophical significance of the facts of nature. To introduce another term, science compels us to explore "metaphysical" questions. The

Greek preposition "*meta*" carries the meanings "after," "behind," and "beyond." Therefore, after we have finished our scientific investigations, we inevitably think about metaphysics and ultimate beliefs that are behind or beyond the physical world.

The relationship between our metaphysical beliefs and scientific discoveries is a two-way exchange of ideas as depicted by the upward and downward arrows in Figure 3–2. Most notably, this complementary relationship between metaphysics and physics is rooted in an act of faith. The reason for this is because there is no direct connection or mathematical formula to move from physics to metaphysics, or vice versa. Consequently, it is crucial to understand that everyone takes a step of faith (or intellectual jump or leap) from science to their religious and philosophical beliefs about nature; a step of faith also occurs from their metaphysics to assumptions that influence their scientific explorations, such as factors that impact their observation of the physical world.

These two reciprocal acts of faith between metaphysics and physics are informed by intuition and reason. Some people arrive quickly at their ultimate beliefs regarding nature and their assumptions undergirding science through an intuitive insight. For others, metaphysical views about physical reality and beliefs that affect scientific practices arise through a slower and more analytical rational process. Most of us use a combination of both intuition and reason in attempting to understand the world and our existence.

To further explain the Metaphysics–Physics Principle, consider the average cell in our body. It's about 1/1000th of an inch wide. If it was placed on the tip of a pin, we wouldn't be able to see it. In this single cell, about two yards of DNA is tightly coiled into chromosomes. The information contained in the DNA of this one cell is approximately similar to the amount of information in 200 telephone books of 1,000 pages each![5] These are scientific facts and no scientist disagrees with them. For example, atheist Richard Dawkins and I are both biologists. If we entered a laboratory to study the average cell, we would arrive at these same facts.

Now here's where it gets interesting. I am sure you are asking the questions, What are we to make of this astonishing scientific evidence about the cell? Does this incredible complexity point to an Intelligent Designer who is a Supreme Engineer? For some people such as me, we take a step of faith *upward* from these scientific facts and come to the metaphysical belief that this average cell in our body reflects intelligent design. This physical evidence also leads us to believe that nature features teleology. In contrast others like Richard Dawkins, in making an *upward step of faith*, arrive at the ultimate belief that the cell is not designed and points only to a dysteleological world.

There is also a very subtle *downward* act of faith with the Metaphysics–Physics Principle that is often overlooked and rarely acknowledged. Our ultimate beliefs act like a "metaphysical filter" through which we observe and interpret the natural world. Let me explain this notion with two examples.

Religious individuals believe in a Creator and intelligent design. When exploring nature through science, we look at the universe through the eyes of religious belief and see greater manifestations of design and teleology than anti-religious individuals do. It is important to note that Richard Dawkins also takes a *downward step of faith*, but he examines the physical world through the eyes of atheistic belief. This metaphysical assumption leads him and other atheists to ignore and dismiss any reflections of teleology and intelligent design. In fact, their atheistic beliefs are factors that often drive their scientific investigations in attempting to discover features in nature that purportedly point to a dysteleological universe.

As noted previously, there is no direct connection or mathematical formula that allows us to move from science to ultimate beliefs, or from ultimate beliefs to science. This is the case because metaphysical reality and physical reality are radically different realms. Therefore, whether people are aware of it or not, or whether they believe in God or not, we are all forced to take a step of faith in understanding the relationship between scientific discoveries and our metaphysical beliefs.

The Metaphysics-Physics Principle underlines that everyone has a personal faith, including anti-religious individuals like Richard Dawkins. It is only through an act of faith that an atheist comes to the *belief* that there is no design, no teleology, and no Creator. In the next chapter we will explore the notion of intelligent design in more detail. I will propose the provocative idea that human sinfulness is a significant factor that impacts the step of faith involved in assessing the existence of design in nature.

—Scientific Concordism and Spiritual Correspondence—

In the last chapter, I introduced the terms "concordism" and "scientific concordism." The former sometimes appears in books on biblical interpretation and the origins debate. I used the latter term to emphasize the issue of whether modern science appears in the Bible. Scientific concordism is such an important idea that I will ask you to think about it again. In order to understand the various Christian views of origins, we need to identify the assumptions that each position makes about the relationship between Scripture and science.

Scientific concordism is the idea that God revealed some basic scientific facts in the Bible thousands of years before their discovery by modern science. As the lower half of Figure 3–3 presents, scientific concordism assumes that statements about nature in Scripture match up with the facts of physical reality. As I stated before, I believe that God is powerful and that he transcends the creation and time. Since he is both the Creator of the world and the Author of the Bible, it is very reasonable for Christians to expect some sort of alignment between the Book of God's Words and the Book of God's Works. But the question arises, do Scripture and science line up? This is a critical question that you need to answer in developing your view of origins.

The upper half of Figure 3–3 introduces another term that is significant to the topic of origins. "Spiritual correspondence" is the belief that statements about spirituality in the Bible align with spiritual reality. I prefer to use the term "correspondence" instead of "concordism" since the latter appears in contexts dealing with physical reality.[6]

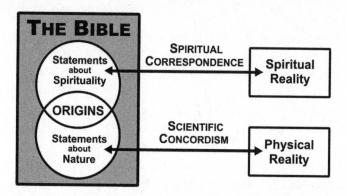

Figure 3–3. Spiritual Correspondence and Scientific Concordism.

Moreover, the word "correspondence" carries the idea of a communication, like a personal letter. Therefore, spiritual correspondence refers to the belief that the personal God of Christianity communicates spiritual truths to men and women through Scripture.

To offer a couple of examples of spiritual correspondence, consider the first chapter of the Bible. Genesis 1:27 asserts that "God created men and women in his own image," and Genesis 1:31 states that "God saw all that he had made, and it was very good." The notions of the Image of God and the goodness of creation are religious beliefs, not scientific facts. They cannot be detected by any scientific instrument. Christians throughout history have believed that these two spiritual statements align with spiritual reality and that they are absolutely true. In other words, these are *inerrant* spiritual truths.

You will have noticed in Figure 3–3 that the issue of origins in the Bible features an overlapping area between statements about spirituality and statements about nature. In my opinion, this is where the greatest challenge appears in attempting to develop our view of origins. To explain what I mean, consider an issue that is often debated among Christians.

Genesis 1 refers ten times to God creating living organisms "according to their kinds." Is this a scientific statement about how he actually made different types of plants, birds, sea creatures, and land animals?

Many Christian anti-evolutionists claim that this phrase is biblical proof that the Creator did not employ evolution to create life. They contend that God used miraculous interventions to create separate "kinds," or groups, of creatures individually (e.g., all dogs as a kind). However, other Christians like me suggest that Genesis 1 does not reveal *how* the Lord created living creatures. Instead, this creation account reveals *who* made plants, animals, and humans—the God of Christianity.

Here are a few ideas I would like you to consider regarding scientific concordism and spiritual correspondence. Is it possible that statements about nature in Scripture do not align with the physical world because God accommodated and allowed the biblical writers to use the science-of-the-day? More specifically, did he communicate timeless spiritual truths by using an ancient understanding of origins as a vessel to deliver them? In other words, is it reasonable to reject scientific concordism, but to accept spiritual correspondence?

Freedom Leads to New Possibilities

As you read this chapter, I hope you experienced something. I hope you felt the power of words. The proper definitions of terms can free us from being trapped in careless conflations and the "either/or" thinking of the origins dichotomy. When we use words accurately, it opens the way to discovering a variety of new possibilities regarding the origin of the universe and life.

For example, there is not just one type of creationist, but many different kinds of creationists. Some creationists believe God made the world in six literal days six thousand years ago. Others claim he formed the universe over long periods of time and intervened at different times to create living organisms. Still others say God created using evolution. There are also various types of evolutionists. Many believe that evolution is dysteleological with no plan, no design, and no God. But there are other evolutionists who argue that evolution is teleological and that the God of Christianity created the evolutionary process. In fact, these Christian evolutionists would even say that evolution reflects design and points back to an Intelligent Designer.

Now that we've defined some basic terms, let's turn to the fascinating topic of intelligent design in nature. I find this issue always strengthens my faith and provides a marvelous example of how Christianity and science can be in a fruitful and complementary relationship.

CHAPTER 4

INTELLIGENT DESIGN AND THE BOOK OF GOD'S WORKS

The natural world is incredibly beautiful, intricately complex, and functions extremely well. Throughout history, the Book of God's Works has deeply impacted every man and woman. This powerful experience has led nearly everyone to believe that the universe and life have been intelligently designed by a Creator who is like a Cosmic Artist and Supreme Engineer.

The belief in intelligent design is not limited to Christians. It is accepted in other religions, appears with various philosophies, and is defended by many world-class scientists. Anyone looking up at distant galaxies through a telescope or down into the cells of our body with a microscope cannot help but ask, "Is there an Intelligent Designer behind this absolutely astonishing world?" The concept of intelligent design continues to be one of the best reasons for believing in the existence of God.

Before we begin, I need to comment again on a view of design that is often promoted in our churches called "Intelligent Design Theory." As noted in the last chapter, ID theorists reject evolution and claim that God intervened miraculously to put design in living organisms. The leading ID theorist is Michael Behe. He coined the term "irreducible complexity." In his famous book *Darwin's Black Box*, he asserts that "if a biological system cannot be produced gradually it would have to arise as an integrated unit, in one fell swoop."[1] To illustrate a structure that is irreducibly complex, Behe points to the flagellum in bacterial cells as sketched in Figure 4–1.[2] He argues that the flagellum did not evolve through natural processes, but was created fully formed.

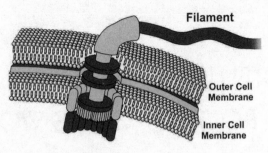

Figure 4–1. Bacterial Flagellum.

No doubt about it, the flagellum is absolutely spectacular! This motor-like structure is embedded in the cell membrane of bacteria. It has a long hair-like filament that spins at about 1,000 revolutions per minute and acts like a propeller to move the cell. For me, the flagellum definitely reflects intelligent design and points to an Intelligent Designer. However, components which make up the flagellum are in the cell membrane and function for different purposes.[3] In other words, these cellular parts were already in existence *before* this structure evolved. The use of pre-existing components to form new structures is a common evolutionary mechanism known as "recruitment." To think that the Creator had the foresight to set up the laws of nature for the flagellum to self-assemble during evolution is mind-boggling! I consider this a stunning example of *evolutionary* intelligent design.

Behe's use of the flagellum is a classic example of what is known as the "God-of-the-gaps." This view of divine action contends that God intervenes miraculously at different times and places in the origin and operation of the physical world to add missing parts or modify existing ones. But the history of science reveals that every time a Christian has proposed a *gap in nature* where God is supposed to have intervened, scientists have later shown that it is a *gap in knowledge* regarding how natural processes work. Therefore, to avoid confusion in our churches, Intelligent Design Theory should be renamed the "*Interventionist* Design Theory" because it is a God-of-the-gaps view of origins requiring divine interventions to produce so-called "irreducibly complex" design.

It is important to qualify that the term "God-of-the-gaps" deals only with divine action in the origin and operation of the universe and life. This concept *does not* include the Lord's miraculous activity in the lives of men and women. Therefore, it is possible for a Christian to reject God-of-the-gaps interventions in the origins and operations of the world, but then to fully embrace personal miracles as well as the miracles of Jesus and his bodily resurrection from the grave. This is my view of divine action.

As noted in the previous chapter, intelligent design is not a scientific theory. It is a religious belief and a metaphysical concept. Precisely defined, intelligent design is the belief that beauty, complexity, and functionality in nature point toward an Intelligent Designer. To fully grasp the notion of design, we need to review the various ways that God reveals himself to us.

Types of Divine Revelation

There are two basic types of divine revelation. *Special Revelation* is specific information from God that is given to men and women, the nation of Israel, and the Christian church. The greatest act of divine revelation is when God became a man in the person of Jesus so we could know him personally and experience his love for us. This revelatory event is known as the "Incarnation." The Latin word *carnis* means "flesh." The gospel of John calls Jesus "the Word" and states, "The Word was God. . . . The Word became flesh and made his dwelling among us" (John 1:1, 14).

Biblical Revelation is another form of special revelation. In particular, this type of divine disclosure is verbal. The Latin noun *verbum* means "word." The Bible contains "the very words of God" (Rom. 3:2) because "all Scripture is God-breathed" (2 Tim. 3:16). This revelation offers specific information about who God is and what his will is for us. Special revelation also includes *Personal Revelation*. In this case, the Lord reveals himself specifically to individuals and to his chosen people. He speaks to us through answering prayers (Mark

11:24), in dreams and visions (Acts 18:9), and with signs and wonders (Heb. 2:4).

The second basic type of divine revelation is *General Revelation*. It is termed "general" because it is experienced by all men and women, including both religious and non-religious individuals. This divine disclosure offers a broad outline of God's attributes and his will for humans. In contrast to special revelation, general revelation is non-verbal. That is, it does not use actual words.

Natural Revelation is a form of general revelation and deals with intelligent design in nature. The creation points to the existence of God and declares his glory (Ps. 19:1). Through beauty, complexity, and functionality, the natural world offers some general attributes of the Creator, such as his divine nature and eternal power (Rom. 1:20). General revelation also includes *Moral Revelation*. Often referred to as our "conscience," God has "written on [the human] heart" a revelation that guides us in understanding right from wrong (Rom. 2:14–15).

Christians throughout the ages have believed that the Lord reveals himself through both Scripture and nature. They have often used the metaphor of God's Two Books to depict these divine revelations.

The Book of God's Words provides a verbal and specific revelation. For example, Scripture offers some of the Lord's personal and special attributes. God is holy (Rev. 4:8) and he is love (1 John 4:8). He is also just (Isa. 30:18) and merciful (Deut. 4:31). In addition, the Bible discloses that humans were made in the Image of God (Gen. 1:27), the world is a very good creation (Gen. 1:31), and the Creator of the world is specifically the God of Christianity (Gen. 1:1; John 1:1–3).

The Book of God's Works is a non-verbal and general revelation. The many artistic and engineered features in nature are reflections of intelligent design and point to an Intelligent Designer. Through the study of science, we are offered an outline of God's general attributes such as his glory, power, and divinity. In fact, the history of science demonstrates that as scientists have probed deeper into nature through microscopes and telescopes, we have enjoyed greater and more magnificent revelations of the Creator.

BOOK OF GOD'S WORDS
Verbal Revelation
Uses Words
Inspired Writings in Scripture
All God-Breathed
Very Words of God
Specific Attributes of God
Holy, Love, Just & Merciful

BOOK OF GOD'S WORKS
Non-Verbal Revelation
Does **NOT** Use Words
Intelligent Design in Nature
Beauty, Complexity & Functionality
General Attributes of God
Divine, Glorious & Eternally Powerful

Figure 4–2. Divine Revelation and God's Two Books.

Figure 4–2 summarizes divine revelation from the perspective of God's Two Books. As we will see in this chapter, the verbal disclosure in Scripture and the non-verbal revelation in nature complement each other and provide us with an integrated understanding of intelligent design and the Intelligent Designer.

The Bible and Intelligent Design

It is necessary to point out that the terms "natural revelation" and "intelligent design" do not appear in the Bible. However, the idea that nature reveals the Creator through his creation is definitely in the Word of God. The two central passages in Scripture that deal with the concept of intelligent design are Psalm 19:1–4 and Romans 1:18–23. Other notable passages affirming design in the physical world include Proverbs 8:22–31; Job 38–41; Psalms 8:1–9; 104:1–35; 139:13–14; 148:1–14; Acts 14:15–17; 17:22–31. Let's look closely at Psalm 19 and Romans 1.

Psalm 19: The Heavens Declare the Glory of God

Psalm 19 features two basic parts that could be entitled "The Book of God's Works" in verses 1–6 and "The Book of God's Words" in verses 7–11. The first half of this psalm is a rich source of spiritual truths about the divine revelation in nature.

> ¹The heavens declare the glory of God;
>> the firmament proclaims the work of his hands.
> ²Day after day they pour forth speech;
>> night after night they reveal knowledge.
> ³They have no speech, they use no words;
>> no sound is heard from them.
> ⁴Yet their voice goes out into all the earth,
>> their words to the ends of the world.
> In the heavens God has pitched a tent for the sun.
>> ⁵It is like a bridegroom coming out of his chamber,
>> like a champion rejoicing to run his course.
> ⁶It rises at one end of the heavens
>> and makes its circuit to the other;
>> nothing is deprived of its warmth.⁴

This remarkable passage has characteristics that are consistent with the notions of natural revelation and intelligent design. Six biblical design categories include:

1. The creation is *active*. Notice the active verbs. These are verbs that describe the actions of someone or something. This short passage uses five of them to emphasize that the physical world thrusts itself upon us. The heavens "declare," the firmament "proclaims," both of these structures "pour forth" and "reveal," and their voice "goes out."

2. The message in nature is *understandable*. The psalm uses four nouns that are associated with intelligent communication: "speech," "knowledge," "voice," and "words." The world was made in such a way that its design displays rationality and the mind of a Creator.

3. Natural revelation is *non-verbal*. Verse 3 states that the heavens "have no speech, they use no words; no sound is heard from them." Yet verse 4 immediately adds that there is a "voice" and "words" that go out into the world. This is to say that the message revealing the glory and work of God is non-verbal. Natural revelation is like music. It does not use actual words, but like a splendid symphony it speaks clearly to everyone.

4. The creation's message is *never-ending*. The non-verbal "speech," "knowledge," "voice," and "words" in nature are heard "day after day" and "night after night" throughout time. Intelligent design in nature has existed from the beginning of the world.

5. Natural revelation is *universal*. The "voice" of this divine message travels "into all the earth" and "to the ends of the world." This disclosure in creation is not limited only to Christians and other religious people. These "words" are heard by every person in every place on earth.

6. The message through nature is a *divine revelation*. It is authored by God and it is about him and his creation. Without using actual words, the "voice" in the physical world "declares the glory of God" and "proclaims the work of his hands."

Psalm 19 also offers an insight into biblical interpretation. The writer uses the ancient understanding of the structure of the world that we noted in Figure 2–1 on page 29. He refers to "the firmament," "the ends of the world," the heavens being structured like a "tent" with a flat floor and domed canopy, and the daily movement of the sun that "rises at one end of the heavens and makes its circuit to the other."

I am sure you will agree that the central spiritual truth in Psalm 19 is that God reveals himself through the creation. Knowing the actual structure and operation of the world is not essential for believing in intelligent design. The biblical notion of design focuses on the belief that nature reflects design, and not on *how* the natural world is actually structured or *how* it operates.

Romans 1: Men and Women Are Without Excuse

Romans 1:18–23 also affirms the reality of God's revelation in nature. In the first part of this passage, the apostle Paul writes:

> [18]The wrath of God is being revealed from heaven against all the godlessness and wickedness of men and women, who suppress the truth by their wickedness, [19]since what may be known about God is plain to them, because God has made it plain to them. [20]For since the creation of the world God's invisible qualities—his eternal power and divine nature—have been clearly seen, being understood from what has been made, so that men and women are without excuse.[5]

Romans 1:18–20 has a number of features dealing with natural revelation and intelligent design that are similar to the six biblical design categories in Psalm 19.

1. The creation is *active*. The impact of "what has been made" by the Creator is so powerful that we "are without excuse" regarding the clear implications of this divine revelation in the physical world.

2. The message in nature is *understandable*. Paul uses words that are associated with intelligent communication to describe natural revelation: "known," "seen," and "understood." Nature reflects intelligence because God has built design and rationality into his creation.

3. Natural revelation is *non-verbal*. Though Romans 1 is not as explicit as Psalm 19 regarding this characteristic, it is certainly implied. For example, the wordless creation discloses some of God's "invisible qualities."

4. The creation's message is *never-ending*. It has been "seen" and "understood" ever "since the creation of the world." In other words, this divine disclosure in nature has existed from the beginning of time.

5. Natural revelation is *universal*. Even though it does not use actual words, it has been "made plain" to everyone and it is "clearly seen" by everyone. The message in creation is not limited to just Christians and other religious individuals.

6. The message in nature is a *divine revelation*. The physical world is like a book in which God has written out "his eternal power and divine nature."

However, Romans 1 goes further than Psalm 19 and includes two more design categories regarding God's revelation in nature.

7. The message inscribed in the creation is *rejectable*. Even though this divine revelation is "plain" and "clearly seen," God has given humans the freedom to ignore it, reject it, or even to call it an illusion. But there are serious consequences for those who refuse to accept this revelation.

8. The creation makes all men and women *accountable*. The perfectly understandable "voice" in nature puts us in a position where we are "without excuse" regarding its profound implications. There is no justification whatsoever for "godlessness and wickedness" or for anyone to "suppress the truth." The creation is a constant non-verbal reminder declaring to everyone the existence of an eternally powerful and divine Creator.

The apostle Paul also contends that natural revelation and intelligent design are closely connected to the spiritual state of a person, in particular, to their sinfulness. He continues in the second part of Romans 1:18–23, starting with verse 21:

> [21]For although they knew God, they neither glorified him as God nor gave thanks to him, but their thinking became futile and their foolish hearts were darkened. [22]Although they claimed to be wise, they became fools [23]and exchanged the glory of the immortal God for images made to look like a mortal human being and birds and animals and reptiles.

Paul is pointing back to the first two commands of the Ten Commandments. The first commandment states, "You shall have no other gods before me;" and the second commandment adds, "You shall not make for yourself an idol" (Ex. 20:3–4; Deut. 5:7–8 NRSV).

In light of Romans 1:21–23, it is evident that by breaking these first two commandments, human sinfulness leads to intellectual dysfunction. According to Paul, those who are ungrateful to God and who embrace idols "became fools" and "their thinking became futile." In other words, *sin impacts our ability to think clearly and rationally.* Paul adds in Romans 1:25 that if we replace God with idols, then we will be "exchang[ing] the truth about God for a lie." Stated another way, in disobeying the first and second commandments, sinfulness twists our thought processes into believing falsehoods.

Romans 1:18–23 affirms the reality of natural revelation and intelligent design. The divine disclosure in creation is so clear that "men and women are without excuse" if they reject it. Yet some people today argue that design in nature is only an illusion. But for those who say this, is sin a factor in coming to this belief? Could it be that their argument against design is the result of the intellectual dysfunction caused by breaking the first and second commandments? Or, to recast the words of the apostle Paul, are those who claim intelligent design is a figment of human imagination "exchanging the truth about God for a lie?"

To be sure, this is a provocative idea. It is important to emphasize that I am in no way judging anyone—only God can do that. But I believe that sin impacts our ability to think. And our spiritual state influences our belief about intelligent design.

Nature and Intelligent Design in a Complementary Relationship

Now that we have examined the two most important passages in Scripture related to the notion of intelligent design and have identified eight biblical design categories, let me propose a way to understand design in nature. This approach is structured on the Metaphysics-Physics Principle as shown in Figure 4–3. I believe that there is a complementary relationship between the ultimate (metaphysical) belief in intelligent design and the scientific (physical) discoveries in nature. In particular, reciprocal steps of faith are essential components in the various interpretations of design.

Together Psalm 19:1–6 and Romans 1:18–23 assert that the

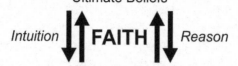

Figure 4–3. A Complementary Relationship between Nature and Intelligent Design.

creation offers a divine message that is active, understandable, nonverbal, never ending, universal, revelatory, rejectable, and makes humans accountable. This natural revelation is manifested through a wide variety of artistic and engineered features seen throughout the physical world. Of course, nature strikes different people in different ways. Some are moved more by the breathtaking beauty, others by the machine-like structures (e.g., the flagellum). Most people experience a combination of both and view the world as reflecting the rational designs of an Intelligent Designer who is a Cosmic Artist and Supreme Engineer.

The reciprocal arrows in Figure 4–3 represent the two-way exchange of information between the religious/philosophical belief in intelligent design and the scientific discoveries in nature. This interchange of ideas occurs with every interpretation of design, including those that reject design. Let me explain.

First, science offers countless examples of nature's stunning beauty, sophisticated complexity, and astonishing functionality. By taking a step of faith *upward* from this physical evidence, we can come to the metaphysical belief that these features reflect intelligent design and point to the existence of an Intelligent Designer. This approach is evidential in that it *begins* with evidence in the natural world to argue for the belief of design. I term the use of scientific discoveries in this manner the "argument *from* nature *to* design."[6]

It is necessary to emphasize that atheists like Richard Dawkins also make an *upward* step of faith from scientific evidence to arrive at their secular belief that nature gives only the illusion of being designed by a Creator. And this is a belief, no different than a religious belief, because it is not a scientific fact. This is why some people view atheism as a secular religion.

Second, religion and philosophy provide the metaphysical belief in design through which we see and even expect reflections of intelligence in nature. This time, through a *downward* act of faith, we offer science the ultimate explanation for the existence of artistic and engineered characteristics in nature: these features are creations of an Intelligent Designer. Our faith in design also acts like a metaphysical filter that influences our observation of the physical world in that we see more manifestations of intelligence than those who reject design. This approach is presuppositional in that it *begins* with the assumption that intelligent design exists, and then it views the creation through this lens to discover reflections of design in nature. I call this way of reasoning the "argument *from* design *to* nature."

It must be underlined that atheists such as Dawkins make a similar *downward* step of faith. They take their secular belief rejecting intelligent design and view nature through it. In other words, by observing the world through an atheistic metaphysical filter, they attempt to explain away the reflections of design as being merely illusions in the mind.

An act of faith is an indispensable component in every interpretation of intelligent design, as depicted in the center of the diagram in Figure 4–3. One of the best verses in Scripture that explains the role of faith in understanding the ultimate meaning of the natural world comes from Hebrews 11, the great biblical chapter on faith. Hebrews 11:3 states, "By faith we understand that the universe was formed at God's command, so that what is seen was not made out of what was visible."

To recast this verse within the context of intelligent design, Christians "by faith" use scientific discoveries to "understand" that the world is intelligently designed by a Creator. Moreover, by believing in design and looking at the universe through the eyes of faith, we "understand" that the beauty, complexity, and functionality in nature are the work of an Intelligent Designer. Christians undoubtedly perceive more reflections of design than are perceived by the eyes of atheists who are obstructed by their faith that design is just an illusion. Therefore, our belief in intelligent design features a complementary relationship between the argument *from* nature *to* design and the argument *from* design *to* nature.

Let's consider this reciprocal relationship between nature and intelligent design by using the notion of God's Two Books. On one hand, the Book of God's Works, through its incalculable artistic and engineered characteristics, points to the "glory," "eternal power," and "divine nature" of an Intelligent Designer. In this way, the physical world affirms the biblical revelations in Psalm 19:1 and Romans 1:20. On the other hand, the Book of God's Words in Psalm 19:1–6 and Romans 1:18–23 reveals that we should expect a divine message in the creation that is active, understandable, non-verbal, never-ending, universal, revelatory, rejectable, and makes humans accountable. From this perspective, Scripture explains and elaborates on the revelation in nature and confirms that the Intelligent Designer is the God of Christianity.

My approach to understanding design also acknowledges human accountability. The apostle Paul claims in Romans 1:20 that we are "without excuse" if we reject the divine revelation in nature pointing to God's existence and some of his attributes. Coupling this belief in our accountability with the faith needed to believe the world was formed by God (Heb. 11:3) suggests that our knowledge of design is not at the level of a scientific or mathematical proof. Instead, it is similar to a powerful argument. To borrow a term used by lawyers, the level of certainty regarding the existence of intelligent design in nature is "beyond a reasonable doubt."

Finally, my view of design recognizes human sinfulness as a critical factor that influences our interpretation of intelligent design. In Romans 1:21–22 the apostle Paul asserts that sin impacts the mind and causes intellectual dysfunction. Those who do not give thanks to God and replace him with idols become "fools" and "their thinking" becomes "futile." I am completely unapologetic for my belief that breaking the first and second commandments plays a crucial role in the rejection of intelligent design. Sin affects human intuition and reason and ultimately impacts the step of faith taken to believe that design in nature is merely an illusion.

Is Intelligent Design an Illusion? A Response to Richard Dawkins

Richard Dawkins is obsessed with design in nature. He wrote a 350-page book attempting to write off intelligent design as an illusion in our minds. It is odd that any atheist would take so much time and effort to do so. But maybe Dawkins hears the "voice" of a non-verbal divine revelation in the physical world, and as an atheist he needs to find a way to justify his rejection of its clear message.

The Blind Watchmaker is Dawkins's well-known book attacking intelligent design. In the opening pages, he makes an astonishing confession:

> The problem is that of complex design. . . . The complexity of living organisms is matched by the elegant efficiency of the *apparent* design. If anyone doesn't agree that this amount of complex design cries out for an explanation, I give up. . . .
>
> Our world is dominated by feats of engineering and works of art. We are entirely accustomed to the idea that complex elegance is an indicator of premeditated, crafted design. This is probably the most powerful reason for the belief, held by the vast majority of people that have ever lived, in some kind of supernatural deity. . . . It is almost as if the human brain were specifically designed to misunderstand Darwinism [atheistic evolution], and find it hard to believe.[7]

Let's examine this passage in detail. Remarkably, it points back to nearly all the biblical design categories in Psalm 19 and Romans 1.

First and foremost, Dawkins admits that nature powerfully impacts everyone, *including himself*. He confesses that "complex elegance" in the world hits him so hard that it creates a "problem." Unwittingly, Dawkins claims that "complex design *cries out* for an explanation" (my italics). This sounds quite similar to Psalm 19:1: "The heavens *declare* the glory of God." Dawkins's personal experience of the physical world confirms that the creation is active and thrusts itself upon us through its design.

Second, Dawkins acknowledges that this experience in nature leads to the notion of God's existence. He also agrees that the impact of creation is universal and never-ending in that nearly everyone throughout history has understood this message perfectly. Dawkins states that "for the vast majority of people that have ever lived," intelligent design was "probably the most powerful reason" for their belief in a "supernatural deity." I think Dawkins is absolutely correct. The fact that he is obsessed with trying to destroy belief in design is proof he completely understands the message in nature declaring the reality of a Creator.

Third, Dawkins agrees that artistic and engineered features in the world are an "indicator of premeditated crafted design." Note that these "feats of engineering" and "works of art" are non-verbal. Dawkins also sees a wonderful balance between these characteristics as evident with his use of the terms "elegant efficiency" and "complex elegance." For me, this is the best idea in the Dawkins passage. Intelligent design is not limited to complexity and functionality. It also includes beauty. God is not just a Supreme Engineer; he is also a Cosmic Artist.

Finally, Dawkins rejects intelligent design and believes it is nothing but an appearance. In other words, design is merely an illusion in the minds of most people.[8] He then concludes, "It is almost as if the human brain were specifically designed to misunderstand Darwinism [atheistic evolution], and find it hard to believe." For the moment, disregard Dawkins's misrepresentation of Darwin's beliefs. In chapter

8 we will see that Darwin was never an atheist and that he never accepted so-called "Darwinism." Can you see a way to take Dawkins's conclusion and turn it against him?

As a Christian, I would say that "the human brain was specifically designed *by God* to *understand atheistic evolution,* and I find it hard to believe." From my perspective, the Creator has gifted us with a brain so that when we think about an atheistic view of evolution, we naturally and logically come to the conclusion that atheistic evolution could never be true.

Or to recast Dawkins's words another way, I believe "the human brain was specifically designed *by God* to *understand intelligent design in nature,* and find it hard *not* to believe *in an Intelligent Designer.*" This aligns with Romans 1:20 and the spiritual truth that we are accountable and "without excuse" regarding belief in God's existence. The Lord has created our brain to perceive reflections of design and many of his attributes through the creation.

You may have noticed that Dawkins's attack on intelligent design is directed entirely against the argument *from* nature *to* design (upward arrows in Figure 4–3). But as we have noted, there is two-way exchange of ideas between the ultimate belief in intelligent design and the scientific discoveries in nature. Dawkins seems to be oblivious to the fact that he views and interprets nature through an atheistic metaphysical filter. He thrusts his dysteleological metaphysics rejecting design upon his investigation of the physical world (downward arrows). Is it possible that his personal rejection of God impacts his mind and forces him into believing that intelligent design is only an illusion?

For example, consider what Dawkins states about God in another of his famous books, *The God Delusion:*

> The God of the Old Testament is arguably the most unpleasant character in all fiction; jealous and proud of it; a petty, unjust, unforgiving control-freak; a vindictive, bloodthirsty ethnic cleanser; a misogynistic, homophobic, racist, infanticidal, genocidal, filicidal, pestilential, megalomaniacal, sadomasochistic, capriciously malevolent bully.[9]

I think everyone will agree that this passage does not reflect calm objectivity and rational thought. It's an emotional and irrational rant. As I see it, the launching of this raging metaphysical belief downward into his scientific examination of nature is a significant factor that causes Dawkins to reject intelligent design. Could there be a problem with the first commandment? I'll let you answer the question.

An Evolutionary Perspective on Intelligent Design

The title of this section will certainly capture the attention of Christians because an evolution vs. intelligent design dichotomy exists within most churches. Many assume they are forced to choose *either* evolution *or* design, but can't accept both. In this section I will present scientific evidence from the evolutionary sciences of cosmology, geology, and biology so we can expand our understanding of intelligent design.

It is important to point out that I see reflections of design in the present structures and operations of the world, such as the bacterial flagellum (Figure 4–1). But I also believe that design was manifested during the past through the amazing evolutionary processes that God used to create the universe and life. I find it absolutely incredible that the Creator ordained and sustained the laws of nature to self-assemble into this astonishing creation, including the flagellum. As a result, and to the surprise of most Christians, I have a wider and greater view of intelligent design than that of the anti-evolutionists.

Let's first consider the Big Bang. Scientists have discovered that forces of nature in this unfathomably gargantuan explosion are finely tuned and delicately balanced. For example, the expanding force of the Big Bang is set perfectly against the force of gravity that tries to pull everything back together. If the explosive force had been a bit weaker, the universe would have recollapsed on itself. If it had been slightly stronger, the galaxies would never have formed.

In *God and the New Physics*, cosmologist Paul Davies notes that the precision between the explosive and gravitational forces is one part in 10^{60}. If we wrote out this number, it is 1 in 1,000,000,000,000, 000,000,000,000,000,000,000,000,000,000,000,000,000,000,000.

Davies explains, "To give meaning to this number, suppose you wanted to fire a bullet at a one-inch target on the other side of the universe, 20 billion light-years away [120,000,000,000,000,000,000, 000 miles]. Your aim would have to be accurate to that same part in 10^{60}."[10] And this is just one calculation between only two forces in the Big Bang!

Mathematician Roger Penrose in his *The Emperor's New Mind* offers an estimate of the amount of precise order at the Big Bang.[11] He calculates it to be a mind-boggling one part in $10^{10^{123}}$. In attempting to understand Penrose's estimation, let's consider another number, 10^{10^2}. As you know, 10^2 is 100. Therefore, 10^{10^2} can be re-written as 10^{100}. If fully written out, this is a 1 with 100 zeros behind it. To give you an idea of how incredibly large this number happens to be, it has been estimated there are approximately 10^{80} atoms in the *entire* universe.

Continuing on as above, 10^3 is 1000 and so 10^{10^3} is 10^{1000} which is a 1 followed by 1000 zeros. 10^4 is 10,000 and so 10^{10^4} is $10^{10,000}$ which is a 1 followed by 10,000 zeros, and so on and so forth. Now let's return to Penrose's calculation of $10^{10^{123}}$. This number is $10^{1,000,000,000,000,000,000,}$ $_{000,000,000,000,000,000,000,000,000,000,000,000,000,000,000,000,000,000,000,000,000,000,000,000,000,000,000,000,}$ $_{000,000,000,000,000,000,000,000}$ which is a 1 followed by 10^{123} zeros. This number is so unfathomably gigantic that it could never be fully written out by anyone during their lifetime. In other words, it is utterly impossible for the human mind to grasp the overwhelming precision of the Big Bang. From my perspective, this mind-numbing scientific evidence points to a Creator who fine-tuned cosmological evolution.

Turn our attention now to planet Earth. In their book *Rare Earth*, geologist Peter Ward and astronomer Donald Brownlee challenge the common assumption that the earth is just one of a countless number of planets in the universe that can support advanced life. They argue convincingly that the "nearly perfect habitat" of Earth seems to be "extraordinarily rare."[12] Ward and Brownlee offer a wide range of scientific facts that indicate our planet has the "right" kind of characteristics for life to evolve.[13]

For example, the Earth is in the "right kind of galaxy" that allows the evolution of enough metals for life to exist. Plants and animals require iron, zinc, and copper in a variety of biological processes. The metal core of our planet produces magnetic fields necessary to deflect life-damaging radiation from outer space. And radioactive metals provided an internal source of heat in the Earth that caused volcanoes to release water and gases, resulting in the evolution of the oceans and atmosphere.

The Earth is also the "right distance" from the sun and allows for liquid water, which is absolutely necessary in the evolution of life. The moon is the "right distance" from the Earth and stabilizes the "right tilt" so that seasons are not too severe. Our Earth is the "right planetary mass" to retain an atmosphere, but allows harmful gases to escape. Similarly, the sun is the "right mass" in that it does not emit too much life-harming ultraviolet radiation. The Earth also has the "right amount" of carbon. This molecule is the backbone of the complex molecules in living organisms. There is not too much carbon causing excessive heat and a runaway greenhouse effect, and not too little inhibiting the evolution of life.

In evolutionary biology, paleontologist Simon Conway Morris proposes that the laws of nature were programmed for life to evolve. In *Life's Solution*, he challenges the popular idea in biology that evolution is completely random and without any direction. According to that dysteleological belief, if the evolutionary process were to be started again, it would not produce the plants and animals we have today. It might not even produce life.[14] But Conway Morris believes evolution is teleological and directed toward a final goal. He contends that repeating the evolutionary process would create basically the same living organisms, including humans.

Conway Morris's central argument is based on a pattern that reappears throughout the fossil record. Nearly identical creatures and features have evolved *independently* from each other on *separate and unrelated* branches of the evolutionary tree. This is known as "convergent evolution." For example, this pattern is found between marsupial animals (they give birth to undeveloped young in pouches) and placental animals (like us who are nourished through an umbilical cord in the womb) as they evolved on different continents. Remarkably similar organisms emerged in both evolutionary lines, such as mice, burrowing moles, long-nosed anteaters, saber-toothed cats, and gliding animals like sugar gliders and flying squirrels.

One of the most striking cases of evolutionary convergence involves the eye. It has evolved independently over forty times across a wide range of different animals, such as insects, molluscs, and vertebrates.[15] Notably, a camera eye like ours and that of the octopus evolved separately six different times. Conway Morris also lists over 400 other striking examples of convergent evolution.[16] This recurrent tendency in biological evolution to create similar organisms and characteristics leads him to believe that living organisms were programmed to evolve. In other words, evolution is not a random process driven by blind chance. Instead, it is intelligently designed to produce specific creatures and features. Conway Morris goes so far as to state that the evolution of humans and our intellectual abilities "is a near-inevitability."[17]

In light of this evidence from the evolutionary sciences, many questions arise. Does the mind-blowing mathematics of the fine-tuning in the Big Bang point to a Fine Tuner who balanced the forces of nature? Is it just a fluke that our planet has all the "right" geological and astronomical features for complex life to evolve? And is it possible that God programmed biological evolution to create plants, animals, and humans?

Of course, it takes a step of faith to either accept or reject the belief that the evolutionary process was designed by a Creator. For me, this evidence from cosmology, geology, and biology not only points to one obvious conclusion but it also strengthens my belief in intelligent design and the God of Christianity. There was a time I thought that the evolutionary sciences never could enrich and deepen the faith of a Christian. But I was wrong.

Some Implications

To close this chapter, I'd like to consider a few implications regarding intelligent design. There is no doubt about it, design in nature is powerful, and it strikes everyone with force, including atheists like Richard Dawkins. However, this natural revelation has limits. Even though the physical world clearly reveals that there is design, it does not tell us precisely *who* the Intelligent Designer is.

The divine revelation in nature is non-verbal. The artistic and engineered features in the creation point to a Creator and offer some of his general attributes. Yet in a subtle way, the Book of God's Works almost calls out for the Lord to reveal himself verbally and more fully. The Book of God's Words does exactly that. The Bible reveals that the Intelligent Designer of this wondrous self-assembling universe and its living organisms is the Christian God.

Another implication of intelligent design is that as science advances, we should expect to discover greater examples of rationality in nature. Indeed this is exactly what the history of science shows us. Through modern telescopes and microscopes, we now see more magnificent manifestations of beauty, complexity, and functionality than ever before, declaring the work of a Cosmic Artist and Supreme Engineer.

Think about the Big Bang and cosmological evolution. It is only in the last fifty years that science has discovered the astounding mathematics of the delicate balance between the forces of nature in this explosive event. Consequently, our generation has a much more powerful argument *from* nature *to* design than earlier generations of Christians. We also have an expanded argument *from* design *to* nature. In viewing this scientific evidence through the eyes of our faith, we have a greater appreciation of the mathematical rationality inscribed in the laws of nature.

Intelligent design is a controversial topic. Everyone knows exactly where design in nature logically leads us, especially atheists like Dawkins. If intelligent design is a reality, then it points toward the existence of a Creator. And if there is a Creator, it is reasonable to believe that he is not only the Lord over the entire universe but he is also the Lord of our life.

Intelligent design thrusts us to the feet of our Maker and forces us to deal with the first and second commandments: "You shall have no other gods before me" and "You shall not make for yourself an idol" (Ex. 20:3–4; Deut. 5:7–8 NRSV). Design puts us in our proper place within the creation—we are the creatures and God is the Creator. The consequences of design are deeply personal. For some the divine "voice" in nature along with "the law written on [the human] hearts" (Rom. 2:15) calls for profound changes in their lifestyle.

In ending I would like to share a *speculation* that I have with regard to the most profound implication of intelligent design in nature. It carries eternal consequences. This is a sobering issue, and I hope that you will think seriously about it.

Like many people today, I believe that after we die, we will stand before God on the Day of Judgment. It will certainly be shocking and dreadful for atheists to meet their Maker. Some will undoubtedly attempt to justify their rejection of him by arguing that they saw no evidence for his existence during their lifetime. However, these individuals might be reminded of the many times when the creation struck them with overwhelming force, because it does "declare the glory of God" and "proclaim the work of his hands" (Ps. 19:1). And then sadly, I suspect that the Creator will judge these men and women as being "without excuse" (Rom. 1:20).

CHAPTER 5

ANCIENT SCIENCE AND THE BOOK OF GOD'S WORDS

When Christians find out that I am both an evolutionist and a Christian, they are quick to challenge me. "If evolution is true, then where in the Bible does it say that God used an evolutionary process?" I am certain that you can see the problem with their question. These Christians are making a huge assumption about the Book of God's Words. They presuppose that scientific concordism is a feature of Scripture. In their mind, the Bible must align with the facts of science. They contend that if God employed evolution as his creative method, then the biblical creation accounts need to state that the universe and life evolved.

I then cautiously ask my challengers, "Is it possible that God allowed the inspired writers of Scripture to use their ancient understanding of nature in the Bible? Or to state this another way, instead of revealing modern scientific facts, did the Lord accommodate by coming down to the level of ancient people and employ the science-of-the-day as a vehicle to reveal life-changing spiritual truths? If this is the case, then Scripture has an ancient view of origins."

But these Christians are even quicker to respond. "Are you saying that God *lied* in the Bible?" Challenging their belief in scientific concordism is usually interpreted as an attack on the Word of God, the Christian faith, and God himself. The reason they react in this way is because they conflate concordism and Christianity. I immediately assure them that God does *not* lie in the Bible. In fact, Scripture states directly in Titus 1:2 that God "does not lie" and Hebrews 6:18 asserts that "it is impossible for God to lie." Lying requires an individual to be deceptive, and the God of the Bible is not a God of deception.

Questioning the truthfulness of scientific concordism is difficult. This assumption is deeply ingrained within us by our churches and Sunday schools, even though we may not be aware of this term. I know this personally. However, in my theology training, my professors introduced me to the ancient science in Scripture. After many struggles, I slowly came to accept the idea that God had accommodated in the Bible and permitted the use of an ancient understanding of origins in the creation accounts.

There was a critical moment when all this biblical evidence pointing to an ancient science came together and exploded in my mind. It was a terrifying and dark moment. For about twenty to thirty seconds, I thought about walking away from Christianity. I felt completely betrayed. Why had no one in my church or Sunday school ever taught me about the ancient science in Scripture or the Principle of Accommodation?

Yet God's love and peace quickly welled up inside of me. He made me realize that I was stepping away from scientific concordism, not from him. The Lord assured me that the Bible was indeed the Word of God. It then became clear that the central purpose of Scripture was to reveal spiritual truths about God, his creation, and humans because God wants to be in a personal relationship with each of us.

Reading the Bible through Ancient Eyes and an Ancient Mind-Set

One of the greatest lessons I learned during my theological education was to read Scripture like a person in ancient times. Though the Bible was written *for* everyone in every generation, it was written *to* a specific ancient people during a specific ancient period.[1]

Let's look at a couple of biblical verses to explain what I mean. Scripture refers to the movement of the sun across the sky. Ecclesiastes 1:5 states, "The sun rises and the sun sets, and hurries back to where it rises." Psalm 19:6 asserts, "[The sun] rises at one end of the heavens and makes its circuit to the other."

Most Christians interpret these verses by claiming the biblical authors used phenomenological language. The Greek noun *phainōmenon* means "appearance." In other words, the so-called "rising" or "setting" of the sun is only a visual effect caused by the rotation of the earth on its axis. It gives us the appearance that the sun "moves" across the sky. But here's the key question we need to ask: *Did the ancient writers of the Bible use phenomenological language in the same way that we use it today?*

History gives us the answer. The idea that the earth rotates daily, causing the visual phenomenon of the sun to rise and set, was accepted only in the seventeenth century, over two thousand years *after* Ecclesiastes 1:5 and Psalm 19:6 were written. Similarly, use of the terms "sunrise" and "sunset" as poetic or figurative language arose only after the scientific discovery of the earth's rotation.

Scripture does employ phenomenological language to describe the natural world. However, there is a subtle and critical difference between what the biblical authors saw and believed to be real in nature and what we see and know to be a scientific fact. For ancient people, observation of the natural world was limited to their unaided physical senses, such as the naked eye. Today we have the privilege of having scientific instruments like telescopes, which have greatly widened and deepened our view of the universe.

Therefore, it is important to recognize that statements in Scripture about nature are from an *ancient* phenomenological perspective. What the biblical writers saw with their eyes, they believed to be real, like the literal rising and the literal setting of the sun every day. In contrast, we view the world from a *modern* phenomenological perspective. When we see the sun "rising" and "setting," we know that it is only an appearance and a visual effect caused by the rotation of the earth on its axis. Figure 5–1 distinguishes between the ancient and modern phenomenological perspectives.

It is crucial that we distinguish these two different phenomenological perspectives of nature. The mistake most Christians make in trying to explain passages that refer to the movement of the sun is that they read these Scriptures through *their* modern phenomenological

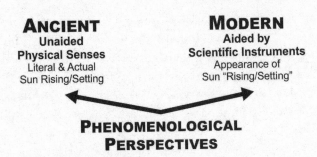

Figure 5–1. Ancient and Modern Phenomenological Perspectives.

perspective. As a result they force *their* modern scientific idea of the visual effect of the sun "rising" and "setting" *into* the Bible. This common mistake is known as "eisegesis." The Greek preposition *eis* means "in, into" and *ēgeomai* is the verb "to guide." The goal of reading is to practice "exegesis." Greek *ek* means "out, out of." We should always attempt to draw *out* the biblical author's intended meaning from a passage, not force a modern idea into it.

Christians must respect that the Book of God's Words was written by God-inspired authors during an ancient period. As Hebrews 1:1 states, "In the past God spoke to our ancestors through the prophets at many times and in various ways." To understand the meaning of biblical passages that deal with the physical world, we need to suspend our twenty-first-century scientific concepts and attempt to think like the people in ancient times who did not have modern instruments like telescopes. In other words, we need to read the Bible through ancient eyes and an ancient mind-set.

The Message-Incident Principle

Let me propose a concept that I find is helpful for interpreting passages in Scripture that refer to the physical world—the Message-Incident Principle as presented in Figure 5–2. It is important to emphasize that this principle is limited to statements in the Bible that deal with nature. *This interpretive concept cannot be used with all statements and every*

Figure 5–2. The Message-Incident Principle.

issue in Scripture. For example, it is not applicable to biblical passages involving the attributes of God (1 John 4:16), his moral commandments (Matt. 22:37–39), or practices in the church such as communion (1 Cor. 11:23–26).

I am convinced that most Christians already accept the basic idea behind the Message-Incident Principle in some implicit way. We all believe that the main purpose of the Bible is to reveal messages of faith and life-changing spiritual truths. I doubt that there are many Christians who go to Scripture to discover scientific facts about nature. For example, no one uses Psalm 19:6 or Ecclesiastes 1:5 as evidence that the sun literally rises and sets every day.

The Message-Incident Principle asserts that spiritual truths in the Bible are *inerrant* because they are absolutely true. Throughout history they have dramatically changed the lives of men and women, bringing us closer to God and providing us joy, comfort, and purpose. These eternal truths are the foundations of Christianity. For example, in Genesis 1 the central messages of faith include God created the universe and life, the world is very good, and only humans have been created in the Image of God. To reveal these truths to the people of ancient times and not to confuse them with modern scientific facts, the Lord accommodated and allowed the biblical authors to employ their ancient understanding of nature. This was the best science-of-the-day, and it was based on an ancient phenomenological perspective of nature.

The ancient science in Scripture is *incidental* because God's central purpose in the Bible is to reveal messages of faith and not scientific facts about his creation. Using the term "incidental" does not mean that ancient science is unimportant. The ancient science in Scripture is essential for transporting spiritual truths. It acts like a cup that holds water. Whether a cup is made of glass, plastic, or metal is incidental. What matters is that a vessel is needed to bring water to a thirsty person. The word "incidental" has the meaning of "happening in connection with something more important." Thus, the incidental ancient science in Scripture is a necessary vessel that delivers important spiritual messages to our thirsty souls.

Let's apply the Message-Incident Principle to one of the most beloved passages in the Bible to see how this principle works. Philippians 2:6–11 reveals that Jesus came down from heaven to become a man and die on the Cross for us. Verses 9–11 state, "Therefore God exalted [Jesus] to the highest place and gave him the name that is above every name, that at the name of Jesus every knee should bow, *in heaven* and *on earth* and *under the earth*, and every tongue acknowledge that Jesus Christ is Lord" (my italics).

Most Christians rarely think about the meaning of the phrase "under the earth." Yet if we examine the original Greek word in Scripture, it is *katachthoniōn*. This noun is made up of the preposition *kata*, meaning "down," and the noun *chthonios*, referring to the "underworld" or "subterranean world." A more accurate translation of Philippians 2:10 would be "that at the name of Jesus every knee should bow, [1] in heaven and [2] on earth and [3] down in the underworld." In other words, this verse is referring to an ancient understanding of the structure of the world known as the "3-tier universe," which is depicted in Figure 5–3.

What are we to make of Philippians 2:10? Does this verse weaken our belief that the Bible really is the Word of God? Or to put it bluntly, did God lie in this verse? My answers to the last two questions are a firm "no" and "no." The purpose of Philippians 2:10 is not to reveal science and the actual structure of the world. Instead, the Message-Incident Principle suggests that the *inerrant* spiritual truth is that Jesus

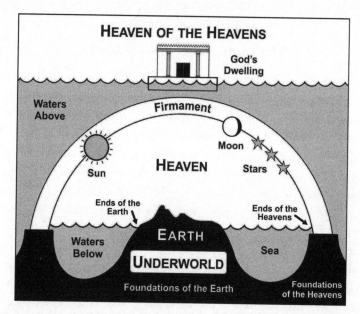

Figure 5–3. The 3-Tier Universe.

is the Lord over the entire creation. In revealing this message of faith to ancient people, God allowed the *incidental* ancient science of the 3-tier universe to be used as a vessel.

To repeat, God did not lie in the Bible. Instead, he accommodated. The power of this eternal spiritual truth in Philippians 2:9–11 is that it can be declared to our twenty-first-century generation by using modern science as an incidental vessel. As Christians, we can proclaim that Jesus is the Lord of our 13.8-billion-year-old universe. Therefore, as science advances, every amazing discovery in nature can be viewed in the light of God's lordship over his creation, including all the discoveries from the evolutionary sciences.

Now that we are equipped with some basic concepts regarding the interpretation of biblical passages that refer to the natural world, let's examine in more detail the ancient science in Scripture. We begin with passages that deal with ancient geography, followed by ancient astronomy, and end with ancient biology.

———————— Ancient Geography ————————

The Immovability of the Earth

Ancient people believed that the earth did not move. This is a reasonable idea from an ancient phenomenological perspective. Even today no one senses that the earth is spinning on its axis at 1,000 miles per hour and moving through space around the sun at 70,000 mph. This phenomenon of not experiencing the earth's movement is so powerful that belief in a stationary earth lasted up until the seventeenth century.

The Bible clearly presents the earth as being immovable. Two verses state directly, "The world is firmly established; it cannot be moved" (1 Chron. 16:30; Ps. 96:10; also Ps. 93:1). Biblical writers often used the engineering term "foundations" to conceptualize the stability of the earth. For example, Psalm 104:5 asserts, "[God] set the earth on its foundations; it can never be moved." But more importantly, this incidental ancient understanding of the earth delivers the inerrant spiritual truth that God is the Creator and Sustainer of our world.

Scripture also refers to a connection between the earth's foundations and the floor of the sea. Psalm 18:15 states, "The valleys of the sea were exposed and the foundations of the earth laid bare at your rebuke, LORD, at the blast of breath from your nostrils" (also 2 Samuel 22:16). This verse make perfect sense in light of the ancient structure of the universe shown in Figure 5–3. Blowing away the sea would reveal the sea floor and the earth's foundations. Again the message of faith is not to reveal the construction of the world, but to emphasize God's lordship over it.

The Circular Earth and the Circumferential Sea

People in the ancient Near East believed that the earth was a circular island surrounded by a circumferential sea. Two phenomenological experiences led to these ideas. First, the horizon gives the appearance

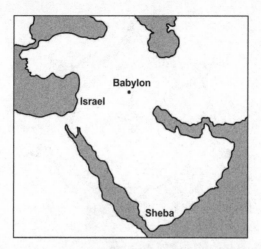

Figure 5–4. Geography of the Ancient Near East.

that the world is enclosed within a circular boundary. Second, travel in the ancient Near East often ended at the shoreline of a sea as shown in Figure 5–4. In fact, the ancient geographical idea of a sea encompassing a circular earth appears in a map of the entire world drawn by the Babylonians around 600 BC, as presented in Figure 5–5.[2]

The ancient notion that the earth had a circular boundary sheds light on part of a verse that is often misinterpreted. Isaiah 40:22 states, "[God] sits enthroned above the circle of the earth." Many Christians claim that this verse refers to the spherical shape of the earth and that it is biblical proof of God revealing a scientific fact in Scripture ahead of its discovery by modern science. In this way, they use the phrase "the circle of the earth" to justify their belief in scientific concordism. However, could this be an example of eisegesis whereby Christians are *forcing* their twenty-first-century understanding of the structure of the earth *into* the Bible?

To answer this question, we first need to read the entire verse. "[God] sits enthroned above the circle of the earth, and its people are like grasshoppers. He stretches out the heavens like a canopy, and spreads them out like a tent to live in." This passage compares the

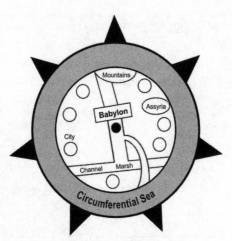

Figure 5–5. Babylonian World Map.

structure of the universe to a tent with a domed canopy overhead and a flat floor below. Thus, Isaiah 40:22 must be understood in the context of a 3-tier universe as pictured in Figure 5–3. The correct interpretation of this verse is that God looks down from his heavenly dwelling and sees the entire earth and its circular boundary meeting the circumferential sea.

Ancient geography also assists us to understand the phrase "the ends of the earth" in Scripture. Ancient Near Eastern people believed that the earth literally came to an end at the shore of the circumferential sea. For example, Jesus proclaims in Matthew 12:42, "The Queen of the South will rise at the judgment with this generation and condemn it; for she came from the ends of the earth to listen to Solomon's wisdom, and now something greater than Solomon is here." The Lord is referring to the Queen of Sheba who lived in the southwest corner of the Arabian Peninsula. From an ancient geographical perspective, her country was at the end of the earth, as seen in Figure 5–4.

The ancient idea of a circumferential sea also sheds light on Proverbs 8:27 and Job 26:10. Both verses point back to the beginning of God's creative activity in Genesis 1:2. "Now the earth was formless and empty, darkness was over the surface of the deep, and the Spirit of

God was hovering over the waters." Proverbs 8:27 states, "[God] drew a circle on the face of the deep" (NRSV), and Job 26:10 says, "[God] marks out the horizon on the face of the waters." In these verses the Creator begins with a flat surface of water upon which he draws a circle to create the horizon.

Most Christians would agree that the purpose of the biblical passages above is not to provide information about the geography of the earth and the sea. Isaiah 40:22, Proverbs 8:27, and Job 26:10 reveal the inerrant spiritual truth that God created the world. Isaiah 40:22 also discloses that he is the Lord over creation. And Jesus in Matthew 12:42 warns that there will be a final judgment. Yet he also encourages us in that he is "greater than Solomon" because he offers us the hope of eternal life.

The Underworld

Ancient people believed in the existence of a subterranean world. The Old Testament refers to this place as *sheōl*. Over half of the times when this word appears, the context and terminology indicate that *sheōl* is below the surface of the earth. For example, God states in Amos 9:2, "Though they dig down to the depths below [of *sheōl*], from there my hand will take them. Though they climb up to the heavens above, from there I will bring them down" (also Ps. 139:8). The expression to "dig down" clearly means tunneling downward through the earth. Amos 9:2 fits perfectly in the context of a 3-tier universe with a subterranean world below and a heavenly realm above.

As noted previously, the Greek noun *katachthonios* means "underworld" in Philippians 2:10. The New Testament also refers to this region as "under the earth." The apostle John in Revelation 5:13 reports, "I heard every creature [1] in heaven and [2] on earth and [3] *under the earth* and on the sea, and all that is in them, saying: 'To him who sits on the throne and to the Lamb be praise and honor and glory and power, for ever and ever!'" (my italics). The Greek noun *hadēs* is another term for the subterranean world. In Matthew 11:23, Jesus declares, "And you, Capernaum, will you be lifted to the heavens? No, you will go down to Hades." Again, these two verses make perfect sense within a 3-tier universe.

For the biblical writers, the underworld was a real place, just like the heavens, the earth, and the sea. But the purpose of the Bible is not to reveal the actual structure of the world. Amos 9:2 offers the spiritual truth that we cannot run or hide from God. Philippians 2:10 and Revelation 5:13 reveal that Jesus is the Lord of the entire universe. And in Matthew 11:23, Jesus gives a stern warning to sinners—there will be a Day of Judgment and accountability.

The Flat Earth

The Bible never states explicitly that the earth is flat. Yet evidence from Scripture presented above indicates that the biblical authors and their readers believed that the earth was like a round coffee table with edges, stable legs, and an underside shelf. Just as when we see the word "earth" and automatically envision a spherical planet, ancient people immediately pictured a flat disk. The idea that the earth was flat is quite logical from an ancient phenomenological perspective. Anyone looking out from an elevated place perceives the world to be a level plain bordered by a circular horizon.

This ancient view of geography is implied in the account of Satan tempting Jesus. Matthew 4:8–9 records, "Again, the devil took [Jesus] to a very high mountain and showed him all the kingdoms of the world and their splendor. 'All this I will give you,' he said, 'if you will bow down and worship me.'" But everyone knows that in Jesus' day there were great civilizations in China and the Americas, and no matter how tall this mountain might have been, it was not possible for him to see all the kingdoms of the world. This biblical passage makes sense only in the context of ancient geography and a flat earth.

Similarly, Revelation 1:7 states, regarding the second coming of Jesus, "'Look, he is coming with the clouds,' and 'every eye will see him, even those who pierced him'; and all the peoples of the earth 'will mourn because of him'" (my italics). Clearly, this verse is best understood if the earth is flat. It is only from an ancient phenomenological perspective that everyone on earth could see the Lord at the end of time.

If God had intended to reveal scientific facts in the Bible, then telling us the shape of the earth would have been easy to do. He could have compared it to something round, like a ball or an apple. The word "earth" appears about 2,500 times in the Old Testament (*'ereṣ*) and 250 times in the New Testament (*gē*). Never once is it referred to as spherical. Surely, if God wanted to reveal scientific facts in Scripture, it would be reasonable to expect him to say something about the structure of the home he made for us. But he never did. This is powerful biblical evidence that scientific concordism is not a feature of the Book of God's Words.

If we apply the Message-Incident Principle to Matthew 4:8–9 and Revelation 1:7, we can avoid a needless conflict between the Bible and modern geography. The purpose of these passages is not to reveal the actual structure of the earth. The geography-of-the-day in Matthew 4 is a vehicle that transports the inerrant spiritual truth that even Jesus was tempted by the Devil, but he did not surrender and bow down to Satan. Instead, he responded with Scripture, "Worship the Lord your God, and serve him only" (Matt. 4:10; see also Deut. 6:13). And in Revelation 1, we are assured that Jesus will come again at the end of time.

Ancient Astronomy

The Firmament

People in the ancient Near East believed that the flat earth and circumferential sea were enclosed by a solid dome. From an ancient phenomenological point of view, the vault of the sky and the circle of the horizon give the appearance of a firm immovable structure overhead, similar to an inverted bowl. As we noted in chapter 2, the Bible refers to this solid heavenly structure as the "firmament." It was created on the second day of creation in Genesis 1 when God separated the "waters above" in heaven from the "waters below" on earth.

The word "firmament" is a translation of the Hebrew noun *rāqîaʿ*. It is derived from *rāqaʿ*, which means "to flatten," "hammer down," and "spread out." This verb carries a sense of flattening something solid. For example, Exodus 39:3 and Isaiah 40:19 use *rāqaʿ* in the context of pounding metals into thin plates. The related noun *riqqûaʿ* refers to metal sheets in Numbers 16:38. And the verb *rāqaʿ* even appears in a passage referring to the creation of the sky, which is understood to be as solid as a metal. Job 37:18 asks, "Can you join [God] in spreading out the skies, hard as a mirror of cast bronze?"

The meaning of the term "firmament" sheds light on biblical passages that refer to the "foundations of the heavens" (2 Sam. 22:8; also Job 26:11) and the "ends of the heavens" (Mark 13:27; also Ps. 19:6). These are logical ideas from an ancient phenomenological point of view. Ancient people would have seen that the firmament was stationary and did not move. Therefore, it was reasonable to believe that it was placed on something solid, like a set of foundations. Similarly, the visual impact of the horizon also led to the conclusion that the dome of heaven came to an end. Figure 5–3 identifies the foundations and ends of the heavens as understood within a 3-tier universe.

Of course, God's purpose in Scripture is not to reveal the actual structure of the heavens. Instead he accommodated to the level of understanding of ancient biblical people and allowed their ancient astronomy to reveal that he alone was the Creator of the heavens. For example, in stating that "the firmament proclaims the work of his hands," Psalm 19:1 (NRSV) employs the ancient astronomical notion of a hard dome as a vessel to carry the inerrant spiritual truth that the heavens reflect intelligent design and point to their Maker.

The Heavenly Body of Water

Ancient Near Eastern people assumed that the firmament held up a body of water in the heavens. Though this idea shocks our twenty-first-century scientific mind-set, such a belief is sensible from an ancient phenomenological perspective. The color of the sky is a changing blue, similar to a lake or sea. And rain falls to the ground from above. These ancient individuals had no way of knowing that the blue of the sky was a visual effect due to the scattering of short wavelength blue light by particles in the atmosphere.

The Bible mentions the existence of a heavenly body of water. As noted previously, on the second creation day in Genesis 1:6–8, the Creator made the firmament to separate the "water above" in heaven from the "water under" on earth (see page 29). In Psalm 104:2–3, "[God] stretches out the heavens like a tent and lays the beams of his upper chambers on their waters." In other words, the Creator sets his divine dwelling on the waters in heaven as depicted in Figure 5–3. Psalm 148:4 calls out, "Praise [the Lord], you highest heavens and you waters above the skies." And Jeremiah 10:12–13 records, "God . . . stretched out the heavens by his understanding. When he thunders, the waters in the heavens roar."

It is worth noting that some Christian concordists argue that the water mentioned in these passages is water vapor. However, the Hebrew language has three well-known words that refer to "vapor" (*nāsî'*, Jer. 10:13 NKJV), "mist" (*'ēd*, Gen. 2:6 NKJV), and "clouds" (*'ānān*, Gen. 9:13). The inspired biblical writers did not use them in the passages cited above. Instead, they employed the Hebrew noun "*mayim*" that refers to "liquid water." This is the same word found in Genesis 1:10 where God names the "gathered waters" as "seas."

The heavenly body of water mentioned in the Bible does not correspond to any physical reality known to modern astronomy. This fact clearly undercuts the common Christian belief in scientific concordism. But this conflict between Scripture and science vanishes in light of the Message-Incident Principle. God accommodated to the intellectual level of the ancient biblical writers and their readers to reveal that he created the huge blue "structure" in the heavens. And this spiritual truth still holds for us today—our Creator made the phenomenon of the blue sky.

The Sun, Moon, and Stars in the Firmament

People in the ancient world believed that heavenly bodies were embedded in the firmament. This is another reasonable idea. The sun, moon, and stars appear to be in front of the blue heavenly body of water and set in the surface of a structure holding these waters up. Scripture presents this ancient phenomenological perspective when God places the "greater light," "lesser light," and "stars" in the firmament on the fourth day of creation in Genesis 1:14–19.

Some Christian concordists claim that the term "firmament" actually refers to "outer space." They know that the sun, moon, and stars are not set in a solid structure. Consequently, they argue that the Hebrew word *rāqîaʿ* should be translated as "expanse" and actually means the vast emptiness of space. Of course, this is eisegesis. These Christians are forcing modern science into the Bible because their belief in scientific concordism drives them to this translation. But as we have seen above, *rāqîaʿ* refers to a hard heavenly structure.

Ancient people also assumed that stars were quite small and that they occasionally dislodged from the firmament and fell to earth. Their appearance as shining specks against the night sky, as well as the sight of a streaking meteor, led to this reasonable idea. For example, in talking about the end of the world, Jesus prophesies in Matthew 24:29 that "the stars will fall from the sky, and the heavenly bodies will be shaken." Similarly, Isaiah 34:4 envisions that "the heavens [will be] rolled up like a scroll; all the starry host will fall like withered leaves from the vine, like shriveled figs from the fig tree."

In the same way that the second and fourth days in Genesis 1 employ an ancient understanding of the creation of the heavens, Matthew 24 and Isaiah 34 use ancient astronomy to describe the heavens being dismantled at the end of time. These passages picture God shaking the firmament and causing stars to fall to earth. Isaiah even views the hard dome of the sky being rolled up. Of course, no one today believes that there is a solid structure above our heads or that all the stars can fall to the earth. Just one star would completely annihilate our planet!

The message of faith in Matthew 24 and Isaiah 34 is not a revelation about the structure of the heavens or exactly how God will disassemble the universe. The inerrant spiritual truth in these passages is that the world will come to an end and there will be a Day of Judgment.

The Lower Heaven and the Upper Heaven

Ancient Near Eastern people believed that the heavens were made up of two basic regions. The lower heaven included the air, the firmament with sun, moon, and stars, and the heavenly waters above. The upper heaven is where God and other heavenly beings resided. It is important to emphasize that both places were real *physical* locations, according to these ancient people. The Bible presents this ancient two-part structure of the heavens.

In the Old Testament, the Hebrew word for "heavens" is *shāmayim*. It has a number of meanings, and the context of the passage usually determines its interpretation. Regarding the lower heaven, this term can refer to the "firmament," when on the second day of creation "God called the firmament Heaven" (Gen. 1:8 NKJV). *Shāmayim* also means "air" as seen in the phrases "every bird of the air" (Gen. 2:19 NKJV) and "the clouds of heaven" (Dan. 7:13). And as previously noted with Psalms 104:3; 148:4; and Jeremiah 10:12–13, the heavens include the heavenly body of water overhead.

The term *shāmayim* also refers to the upper heaven where God is located. The Hebrews prayed to him, "Look down from heaven, your holy dwelling place, and bless your people Israel" (Deut. 26:15). Sometimes the upper heaven is called *shāmayim shāmayim* and translated literally as "heavens of the heavens." This phrase is also rendered the "highest heavens." In praising their Creator, the Israelites proclaimed, "You made [1] the heavens, [2] even the highest heavens, and all their starry host, the earth and all that is on it" (Neh. 9:6). This verse clearly views heaven with two distinct *physical* regions.

In the New Testament, *ouranos* is the Greek word for "heaven." It can refer to the "firmament" since it can be "open[ed]" (John 1:51), "shake[n]" (Heb. 12:26), and "rolled up" (Rev. 6:14). *Ouranos* also means "air" in the phrases "the birds of the air" (Matt. 6:26) and "the clouds of heaven" (Mark 14:62). And *ouranos* is the region where God dwells. For example, "As Jesus was coming up out of the water, he saw heaven being torn open and the Spirit descending on him like a dove.

And a voice came from heaven: 'You are my Son, whom I love; with you I am well pleased'" (Mark 1:10–11).

To summarize, the Bible presents a 3-tier understanding of the structure of the universe as shown in Figure 5–3. This ancient view of the world was the best science-of-the-day thousands of years ago in the ancient Near East. For example, the Egyptians accepted a 3-tier universe as shown in Figure 5–6, and the Mesopotamians understood that heaven had a two-part structure as seen in Figure 5–7.[3] Notably, both diagrams feature a pagan theology with multiple gods. In sharp contrast, Scripture frees the world of these gods and places the God of Christianity as the only Lord over the universe.

Now that we are aware that the Bible includes an ancient view of the structure and operation of the earth and the heavens, it is only logical that there would also be an ancient understanding of living organisms. In other words, ancient geography and ancient astronomy point toward an ancient biology in Scripture.

Figure 5–6. Egyptian Universe. The stars are embedded in the firmament (shaded). The sun god travels in a boat across the heavenly body of water and enters the underworld (lower right corner). The sun passes through the underworld to rise again in the east.

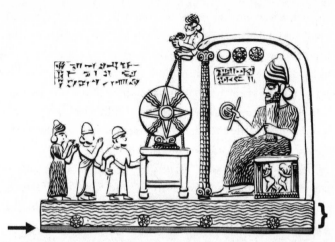

Figure 5–7. Mesopotamian Heavens. The firmament (shaded, arrow) supports the stars and heavenly body of water (bracket). The sun god is seated.

Ancient Biology

Taxonomy

The Bible categorizes living creatures from an ancient phenomenological point of view. For example, the bat is considered to be a bird. Leviticus 11:13–19 states, "These are the birds you are to regard as unclean and not eat because they are unclean: the eagle, the vulture . . . [17 other birds are listed] . . . and the bat." This ancient taxonomical classification is reasonable because bats fly. But we know that bats are mammals. They differ from birds in having body hair and mammary glands, and they develop in their mother's womb and not in an egg with a hard shell.

Similarly, Scripture views rabbits as ruminants like cows. Leviticus 11:6 asserts, "The rabbit, though it chews the cud, does not have a divided hoof; it is unclean for you." The repetitive side-to-side jaw action of rabbits gives the appearance of chewing the cud. However, rabbits are not ruminants and do not have multiple-chambered stomachs.

Ancient taxonomy also appears in Genesis 1. This chapter states ten times that God made plants and animals "according to their kinds." Classifying organisms in this way makes perfect sense. Ancient people would have observed that wheat produced seeds that sprouted only wheat plants. The seeds from various fruit grew into trees that always yielded the same fruit. The ancients also would have seen that hens lay eggs that hatch only chicks, female sheep give birth to lambs, and women are always the mothers of human infants. In the eyes of ancient people, living organisms were *immutable*. In other words, plants and animals never changed and were always the same. Consequently, there is no hint of transitional organisms or biological evolution in the Genesis 1 account of creation.

The purpose of Scripture is not to reveal taxonomical categories. In Leviticus 11 the dietary laws were part of God's plan to separate the Hebrews from the pagan cultures around them to create a holy nation. And in Genesis 1, God accommodated by using an incidental ancient taxonomy of plants and animals to proclaim the inerrant spiritual truth that he alone was the Creator of all living organisms.

Botany

Agriculture and knowledge of plants were critical for the survival of ancient people. Unsurprisingly, the Bible has examples of ancient botany. In the parable of the mustard seed, Jesus asks, "What shall we say the kingdom of God is like, or what parable shall we use to describe it? It is like a mustard seed, which is the smallest of all seeds on earth. Yet when planted, it grows and becomes the largest of all garden plants, with such big branches that the birds can perch in its shade" (Mark 4:30–32).

As most people know, the mustard seed is not "the smallest of all seeds on earth." Orchid seeds are much smaller. Yet from the ancient perspective of the Lord's listeners, mustard seeds were the smallest seeds. Jesus was not lying, he was accommodating. The Lord was descending to the level of understanding of these ancient people and using their knowledge about the size of seeds to deliver a spiritual truth in this parable.

Similarly, Jesus employed an ancient understanding of botany to prophesy his death and resurrection. "The hour has come for the Son of Man to be glorified. Very truly I tell you, unless a kernel of wheat falls to the ground and dies, it remains only a single seed. But if it dies, it produces many seeds" (John 12:23–24; also 1 Cor. 15:36). Seeds do not die before they germinate. If they did, they would decompose and never germinate. However, the outer casing of a seed breaks down and looks like it rots just before germination, giving the perception that a seed dies.

I think that most Christians will agree that Jesus did not come to earth to teach us scientific facts about plants. In the mustard seed parable, he used the ancient perception that this seed was the smallest of all seeds to reveal a prophecy about the kingdom of God. The church would grow out from his small group of disciples and become the greatest of all kingdoms. Likewise in John 12, Jesus was foretelling that through his death and resurrection, many would come to faith in God. And indeed, these two prophecies have come true.

Human Reproduction

Ancient people understood the reproduction of humans in the light of ancient botany. Since agriculture played a major role in their lives, it was natural for them to perceive a similarity between plant seeds and the "seeds" that a man ejaculates during sexual intercourse. It is significant to note that throughout the Bible, only men are said to have reproductive seed, never women.

In the Old Testament, the Hebrew word *zera'* refers to both the "seeds of plants" (Gen. 1:11; 47:23) and the "reproductive seeds of males." Regarding the latter, Leviticus 15:32 in the original Hebrew uses the expression the "flow of seed" to mean the ejaculation of semen. In the New Testament, the Greek noun *sperma* refers to both "plant seed" (Matt. 13:24; Mark 4:31) and "male sexual seed." With regard to ejaculation, Hebrews 11:11–12 in the original Greek states that even though Abraham was a very old man, God "enabled him to throw down seed." In this verse the Greek term *kataballō sperma* means "to ejaculate." As we noted earlier, *kata* is the preposition "down." And *ballō* is the verb "to throw."

This ancient view of human reproduction is known as "preformatism" or the "1-seed model." Ancient people believed that within each male sexual seed there was a tightly packed miniature human. Not having the advantage of microscopes, this was a perfectly logical notion from an ancient phenomenological perspective. Men ejaculate during sex, giving the impression that they are the only contributors of seed in the creation of a life, while women appear to be only receptacles and nurturers of the seed of males.

To illustrate this ancient understanding of reproduction, consider a passage from the Greek writer Aeschylus around 500 BC. "The mother is no parent of that which is called her child, but only the nurse of the new-planted seed that grows. The parent is he who mounts. She preserves a stranger's seed, if no god interferes."[4] Clearly, human reproduction was understood through the 1-seed model, reflected especially with use of the botanical term the "new-planted seed" of the male. Interestingly, preformatism lasted as late as the eighteenth century, even after the invention of the first microscopes, as seen in a depiction of a sperm cell in Figure 5–8.[5]

References to human infertility in Scripture also reflect ancient reproductive biology. As most Christians know, it is only women who are said to be "barren" in the Bible, never men. Well-known examples include Sarah (Gen. 11:30; Heb. 11:11), Rebekah (Gen. 25:21), and Elizabeth (Luke 1:7). In the Old Testament, the Hebrew word translated as "barren" is the adjective *'āqār*. It comes from the verb *'āqar*, which means "to uproot." This term appears in the

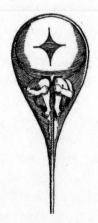

Figure 5–8. Preformatism.

context of agriculture with the phrase "a time to plant and a time to uproot" (Eccl. 3:2). In the New Testament, the Greek word for "barren" is *steira* and it is related to the adjective *stereos* referring to "hard." Why were women in the ancient world considered "uprooted" or "hard" if they could not have children? It is because ancient reproductive biology understood the wombs of women to be like an agricultural field. Barrenness was due to "seeds" and "seedlings" being uprooted after they had been "planted" by a man during intercourse. Seeing miscarriages undoubtedly contributed to this idea. The womb of a barren woman was also perceived to be like hard ground. Ancient people knew that seeds never germinated on compacted earth and grew only in a soft, fertile field. This ancient botany is reflected in Jesus' parable of the sower with seed falling "along the path" and "on good soil," respectively (Matt. 13:4, 8).

In contrast, only a fertile woman bears children, and they are described in agricultural terms. Psalm 127:3 states, "Sons are indeed a heritage from the LORD, the fruit of the womb a reward" (NRSV). And in Luke 1:42, Elizabeth cries out to Mary, who is pregnant with Jesus, "Blessed is the fruit of your womb" (NRSV).

It is necessary to emphasize that passages in Scripture that refer to only females being barren are not intended to disrespect women. This was the reproductive biology-of-the-day. Had we lived at that time, we would have believed this. Of course, modern science reveals that men can also be a source of infertility. But more importantly, these biblical passages show God answering prayer and fulfilling his promises. The miraculous birth of children to barren women is evidence of the Lord's power over nature and his faithfulness to create a holy people to serve his purposes.

Origin of Living Organisms

As we noted above with ancient taxonomy, people in the past were quite reasonable in believing that living creatures were immutable and never changed. In front of their eyes, they saw only that plants and animals reproduce "according to their kinds," and they never observed one type of creature evolving into an entirely different type of creature. With this being the case, let's think about how ancient individuals would have conceptualized the origin of living organisms.

For example, they would have seen that a goat gives birth to a goat, which gives birth to a goat, which gives birth to a goat, etc. In conceiving the origin of goats, they would have reversed this observation of goats giving birth to goats and worked backward through time. In other words, they would have reasoned that a goat today came from an earlier goat, which came from an even earlier goat, etc. And since ancient people believed that living organisms were immutable, they came to the logical conclusion that there must have been an original goat or set of goats created by God.

The conceptual process of reversing the experience of seeing goats giving birth to goats back in time is known as "retrojection." The Latin adverb *retro* means "backwards," and *jactare* is the verb "to cast" or "throw." This is the same type of thinking used in crime-scene investigations. The evidence present at the scene today is cast back in time to re-create criminal events in the past. Similarly, when ancient people reconstructed the period when God created living organisms, they retrojected the series of immutable living organisms giving rise to their own kinds. And they reasonably concluded that the Creator made each plant kind and animal kind quickly and completely formed.

This view of origins is called "*de novo* creation." The Latin preposition *dē* means "from," and *novus* is the adjective "new." It appears in most ancient accounts of creation and features a divine being who acts dramatically through miraculous interventions to make fully formed living organisms.[6] *De novo* (quick and complete) creation was the origins science-of-the-day in the ancient world, and this view was held by the God-inspired authors of Genesis 1 and 2.

To further appreciate ancient understandings of origins, let's examine the origin of humans in some Mesopotamian and Egyptian accounts of creation. There were two basic creative methods. The first was a natural plant-like mechanism whereby humans sprouted out from the earth. The second approach was an artificial craftsman-like fashioning of people using earth or clay.

Regarding the first method, we must remember that the ancients believed in preformatism (1-seed model). The reproductive seeds of men each contained a miniature person. In the *Assur Bilingual Creation Story*, the gods plant the seeds of humans into the earth and people later "sprout from the ground like barley."[7] Similarly, the *Hymn to E'engura* states that "humans broke through the earth's surface like plants."[8] And in the *Song of the Pickaxe*, a god strikes the ground with a hoe-like axe "so that the seed from which people grew could sprout from the field."[9]

Interestingly, this sprouting mechanism seems to be the creative method used in Genesis 1:24 when God commands, "Let the land produce living creatures according to their kinds: the livestock, the creatures that move along the ground, and the wild animals." The Hebrew word translated as "produce" is *yāṣā'*, and it is the same verb found in Genesis 1:12 in which "the land produced vegetation." In this way, Genesis 1 implies that plants and land animals were created from seeds that sprouted out from the earth.

The craftsman method for making humans also appears in various ancient Near Eastern creation accounts. In the *Epic of Atrahasis*, a goddess mixes clay with the blood of a slain god to fashion seven males and seven females.[10] An intoxicated divine being uses earth to make imperfect human beings in the account of *Enki and Ninmah*.[11] The *Epic of Gilgamesh* has a man being created from a pinch of clay.[12] And in Egyptian records, the god Khnum creates people from clay and fashions them on a potter's wheel.[13]

Clearly, these examples of the *de novo* creation of humans are similar to Genesis 2:7 where the Lord God acts like a craftsman and "formed a man from the dust of the ground." In fact, the word translated as "formed" in this verse is the Hebrew verb *yāṣar*, and it is the root of the term "potter." This noun appears in Isaiah 29:16 and 64:8 where God is depicted as a divine potter who forms man in his hands from clay.

To be sure, recognizing that the Bible includes an ancient understanding of the origin of living organisms, including humans, can be quite a challenge to most Christians. The ancient notion of *de novo* creation certainly has significant implications for the origins debate, as we will see in the next chapter. However, the Message-Incident Principle assists us in realizing that the primary purpose of the Bible is to reveal inerrant spiritual truths. The main message in Genesis 1 and 2 is not to tell us *how* God created living organisms, but *who* created plants, animals, and humans—the God of Christianity.

The Bible Is Not a Book of Science

Figure 5–9 offers a summary of ancient scientific ideas in the Bible. It is obvious that concordism fails. Statements about the natural world in Scripture do not align with the facts of physical reality. We do not live in a 3-tier universe. The mustard seed is not the smallest of all seeds. Men do not have miniature preformed people in their semen. Infertility is not limited to women. And land animals do not sprout into existence from the earth. Questioning the truthfulness of scientific concordism is uncomfortable. I know this personally because I was taught in Sunday school that God revealed some basic scientific facts in Scripture.

However, by studying the Book of God's Words in theology school, I slowly came to recognize that the Bible has an ancient understanding of nature. I also realized that if Scripture has an ancient science, I had to acknowledge this biblical fact even if it troubled me. Eventually it became clear that my problem was not with the Word of God itself, but with the Sunday school assumption that scientific concordism is an inerrant feature of the Bible. Like most Christians, I was not aware that I had blindly accepted concordism.

The Lord was good during my theological studies because he sent Christian professors to explain why God allowed the biblical writers to use the science-of-the-day. My instructors emphasized that God did not lie in Scripture. To reveal inerrant spiritual truths, God accommodated and came down to the level of understanding of these inspired ancient authors. If the Lord had stated in the Bible that he

ANCIENT GEOGRAPHY	PHYSICAL REALITY
Earth is immovable	No
Earth is flat & circular	No
Earth has ends & is set on foundations	No
Underworld exists below surface of earth	No
Circumferential sea is flat & surrounds earth	No
Circumferential sea is bordered by ends of heaven	No
ANCIENT ASTRONOMY	
Sun moves across dome of heaven daily	No
Firmament is a solid domed structure overhead	No
Firmament has ends & set on foundations	No
Sun, moon & stars are placed in firmament	No
Stars dislodge from firmament & fall to earth	No
Heavenly body of water is held up by firmament	No
Divine dwelling is set in heavenly body of water	No
ANCIENT BIOLOGY	
Mustard seed is smallest of all seeds	No
Seeds die before germination	No
Bats are birds & rabbits are ruminants	No
Males have reproductive seed	No
Females do not have reproductive seed	No
Womb is like a field for seed of males	No
Infertility is caused only by females	No
Land animals are sprouted from earth	No
Living organisms are created *de novo* (quick & complete)	No

Figure 5–9. Failure of Scientific Concordism.

created through the Big Bang and biological evolution, I doubt anyone in ancient times would have understood. Instead, God used incidental ancient sciences as vehicles to transport life-changing messages of faith. I have personally experienced this as well. My life was not changed by the ancient idea of a 3-tier universe, but rather by the spiritual truth that Jesus is Lord over the entire creation (Phil. 2:10–11).

To close this chapter, consider a few insights from a famous Christian who helped me put the relationship between Scripture and science in perspective. Billy Graham is often regarded as the greatest preacher of the gospel during the twentieth century. In an eye-opening passage he states:

I think we have misinterpreted the Scriptures many times and we've tried to make the Scriptures say things that they weren't meant to say, and I think we have made a mistake by thinking the Bible is a scientific book. *The Bible is not a book of science.* The Bible is a book of redemption, and of course, I accept the Creation story. I believe that God did create the universe. I believe he created man, and whether it came by an evolutionary process and at a certain point he took this person or being and made him a living soul or not, does not change the fact that God did create man. . . . Whichever way God did it makes no difference as to what man is and man's relationship to God.[14]

I love Billy Graham's honesty. He's also correct. Everyone has misinterpreted Scripture. I have made more than my fair share of mistakes, including the mistake of assuming that the Bible is a book of science. But more importantly, Reverend Graham emphasizes that "the Bible is a book of redemption." Scripture convicts us of our sinfulness and reveals that Jesus can restore our relationship with God.

Yet surprisingly, Billy Graham introduces us to an idea that is not often heard in Sunday school. He is open to the possibility that God created humans through evolution. Wow! I was never taught that in my church. How about you? What are your thoughts about the Lord using an evolutionary process to create men and women? And do you agree with Billy Graham that God's creative method does not make any difference with regard to who we are and our relationship with the Lord? No doubt about it, these are not easy questions.

CHAPTER 6

MOVING BEYOND THE "EVOLUTION" VS. "CREATION" DEBATE

The origins dichotomy is alive and well! In February 2014, evolutionist Bill Nye debated creationist Ken Ham. Nye is known as "the science guy" on television and Ham is president of the creation museum in Kentucky. More than 7 million people viewed this so-called "evolution" vs. "creation" debate online.[1] But as we have seen in previous chapters, limiting the topic of origins to only two simple positions indicates that a person is trapped in "either/or" thinking.

In this chapter I will show you that there are more than just two views of origins. We are not forced into choosing between *either* atheistic evolution *or* six day creation. In fact, there are at least five basic positions on the origin of the universe and life: (1) young earth creation, (2) progressive creation, (3) evolutionary creation, (4) deistic evolution, and (5) dysteleological evolution. The first three categories are held by Christians and the last two by non-Christians. This chapter will suggest that everyone needs to get past the misguided origins dichotomy.

It is important to point out that the topic of origins is not limited to only five positions. There are many more views on how the world was made. I have decided to present the five best-known categories so you can start constructing your own personal view. In going through this chapter, you may find you hold to a number of features from the different positions. Selecting and combining these is perfectly acceptable. It means that you will be developing one more view of origins as you move beyond the "evolution" vs. "creation" debate.

At this time, I suggest that you examine the table summarizing the five different categories of origins in Figure 6–1 on pages 126–27. This way you will be able to see where we are heading in this chapter.

—— Positions on the Origin of the Universe and Life ——

Young Earth Creation

Young earth creationists believe God created the entire cosmos and every plant and animal in just six 24-hour days only six thousand years ago. This view of origins is also known as "six day creation" and "creation science." Most people today believe young earth creation is *the* creationist position. It is often seen as *the* official Christian view of origins. As I revealed in chapter 1, there was a time when I believed that *true* Christians had to be young earth creationists.

Six day creation claims Genesis 1 is an accurate scientific description of how the universe and life originated. It contends God created the world quickly and fully developed through dramatic miraculous events. In other words, young earth creationists accept *de novo* creation and firmly reject cosmological evolution, geological evolution, and biological evolution. They insist there is no scientific evidence to support these three evolutionary sciences. Creation scientists argue that *true* science uses Genesis 1 as a guide to interpret the evidence discovered in nature.

Clearly, young earth creationists embrace scientific concordism and read Genesis 1 very literally. They claim that the facts of science line up with the statements in Scripture about the origin of the world. These Christians sincerely believe that the Bible is a book of science.

The most persuasive argument for six day creation is that a strict literal reading of Genesis 1 is the natural and most common way to understand this biblical chapter. A survey of Americans reveals that nearly 90 percent of born-again Christians believe the world was created in one week and Genesis 1 is "literally true, meaning that it happened that way word-for-word."[2]

The majority of Christians throughout history have been young earth creationists. For example, Martin Luther was one of the most famous theologians during the sixteenth century. He argued that the author of Genesis 1 "spoke in the literal sense, not allegorically or figuratively; that is, that the world with all its creatures was created within six days, as the words read."[3] Luther added that Genesis 1 reveals "the world was not in existence before 6000 years ago."[4]

Powerful evidence for a strict literal reading of the Genesis creation accounts comes from Jesus himself. In responding to his critics in Matthew 19:4–5, the Lord asks, "Haven't you read that at the beginning the Creator 'made them male and female,' and said, 'For this reason a man will leave his father and mother and be united to his wife, and the two will become one flesh'?" Jesus was referring directly to Genesis 1:27 and 2:24. Therefore, if any Christian accepts a view of origins other than six day creation, then they need to offer convincing reasons why the opening chapters of Genesis should not be read literally.[5]

A serious problem with young earth creation is that it is in conflict with every modern science that deals with origins. Notably, there are very few six day creationists in the scientific community today. In the United States, a survey reveals that 97 percent of scientists accept that "humans and other living things have evolved over time."[6] Moreover, nearly every university throughout the world fully endorses, teaches, and practices the evolutionary sciences in cosmology, geology, and biology.

This leads to an important question. Are we to believe that all these scientists are completely wrong about evolution? We enjoy the amazing fruits of modern science daily, like medicine and engineering. But are scientists who study the origin of the world totally mistaken? How do we explain that nearly all American scientists are evolutionists? Some Christians say they've been deceived by the Devil. There was a time I believed that. But is that reasonable?

Today the leading young earth creationist in the nation is Ken Ham. He is president of both Answers in Genesis and the creation museum in Kentucky. His most famous book is *Evolution: The Lie*. I had the pleasure of meeting Mr. Ham after church when he visited my city. Though we disagree on origins, he emphasized that understanding how God created the world is not a requirement for being a Christian. He pointed me to Romans 10:9: "If you declare with your mouth, 'Jesus is Lord,' and believe in your heart that God raised him from the dead, you will be saved." I fully agree with Ken Ham regarding our eternal salvation.

Progressive Creation

Progressive creationists accept that the universe is about 13.8 billion years old and the earth 4.5 billion years old. They also believe that God created life sequentially by introducing different living organisms at different points in time. This view of origins is often termed "old earth creation" and "day-age creation." Progressive creation claims that the "days" of creation in Genesis 1 are not 24-hour periods, but actually millions of years long.

According to old earth creation, God used natural processes to create the inanimate universe. He began the world with the Big Bang and then, over billions of years, stars, planets, and moons slowly evolved. Therefore, this view of origins embraces cosmological evolution and geological evolution. However, progressive creationists firmly reject biological evolution. They argue that God intervened miraculously to create various plants and animals during different geological periods.

Progressive creation claims there is a basic alignment between the order of God's acts of creation in Genesis 1 and the appearance of the inanimate world and living organisms discovered by modern science. This origins position accepts scientific concordism and assumes the Bible is a book of science. But in contrast to the concordism of young earth creation, it does not embrace a strict literal reading of Genesis 1. Instead, day-age creationists employ a general literalism and suggest that this biblical chapter offers only a broad outline of divine creative events.

The strongest argument for progressive creation is that it meets a yearning most Christians have for a view of origins based on both

Scripture and science. Every day we are nourished spiritually by the Bible, and we are blessed physically by scientific discoveries. Old earth creation believes that Scripture is the Word of God and it accepts the sciences of cosmology and geology. This origins position offers an approach that integrates Christian faith with these modern sciences.

However, progressive creationists introduce a false dichotomy in science. They uphold the foundational principle that unites the cosmological and geological sciences—the evolution of the inanimate universe. But they dismiss the unifying concept of all the biological sciences—the evolution of living organisms. Consequently, progressive creationists claim that cosmologists and geologists are right in stating that stars, planets, and moons evolved through natural processes, while biologists are completely wrong in claiming that life arose through an evolutionary process. But as noted previously, 97 percent of American scientists today accept the evolution of life, and they would reject a dichotomy between the physical and biological sciences.

Another problem with day-age creation is that it is a God-of-the-gaps view of origins.[7] This understanding of divine activity contends there are gaps in nature where God intervened miraculously. According to progressive creationists, the Creator acted in this way to create living organisms. But the history of science reveals that God-of-the-gaps arguments have always failed. The alleged gaps are gaps in knowledge, not gaps in nature.

The leading progressive creationist today is Dr. Hugh Ross, president of Reasons to Believe. He is a scientist and expert in astronomy. So it is not surprising that he accepts an old universe and cosmological evolution. One of his prominent books is *The Genesis Question: Scientific Advances and the Accuracy of Genesis*.[8] As the subtitle reveals, scientific concordism is the foundational assumption in his view of origins. I have had the pleasure of meeting Dr. Ross and found him to be a remarkable Christian and true gentleman. In my opinion, the Lord has used him mightily in assisting many Christians to understand that our world is billions of years old.

Evolutionary Creation

Evolutionary creationists believe that God created the universe and life through the natural process of evolution. The world did not arise through blind chance, and our existence is not a fluke or mistake. It was the Creator's plan from the beginning to make a world featuring men and women who bear the Image of God. Evolutionary creationists are blessed by the Lord's love and presence in their lives, and they enjoy a personal relationship with Jesus that includes experiencing miracles and answers to prayer.

The term "evolutionary creation" seems like a contradiction in terms. But if we use the proper definitions of terms introduced in chapter 3, it makes perfect sense. The most important word in this origins category is the noun "creation." Evolutionary creationists are first and foremost creationists. They believe in a Creator and that the world is his creation. The adjective "evolutionary" simply indicates the method through which the Lord made the inanimate universe and living organisms.

This origins position is sometimes known as "theistic evolution." Personally, I don't care for this term because it makes the noun "evolution" the more important category; and it turns the Greek noun *theos*, meaning "God," into merely a secondary adjective. I find such an inversion in the priority of words to be completely unacceptable. God is never subordinate to any scientific theory.

Evolutionary creationists believe that the Creator *ordains* and *sustains* all natural processes in the world, including the evolutionary process. This view of origins endorses the sciences of cosmological evolution, geological evolution, and biological evolution. Therefore, these Christians believe that evolution is teleological. It is planned and purposeful, and it has a final goal—the creation of men and women to have a personal relationship with the Lord.

The Embryology-Evolution Analogy helps us to appreciate how evolution can be seen as God's method of creating the world.[9] For example, the DNA in the genes of a fertilized human egg is fully equipped with information necessary for a person to gradually develop in the womb during the nine months of pregnancy. Similarly, the Creator planned the Big Bang and preloaded it with the ability for the universe and life to self-assemble over 13.8 billion years, with humans emerging as the pinnacle of the evolutionary process.

The analogy between embryology and evolution also assists evolutionary creationists to appreciate the appearance of our spiritual characteristics during human evolution. Through our own personal development, each of us began to bear the Image of God, became morally accountable, and then committed sinful acts against our Creator and other humans. In a somewhat similar fashion, prehuman ancestors became fully human when they were given God's Image and made morally responsible. And like each of us, every one of them began to sin.

Understanding exactly how the Image of God and sinfulness arose in our own life seems to be a mystery. For example, I don't remember the very first time I sinned. But not knowing my first sin has no effect on my belief that I am indeed a sinner. This also appears to be the case with the manifestation of human spiritual characteristics in prehumans, resulting in the creation of the first men and women. For me, this is ultimately a mystery and beyond our comprehension. Not understanding how this came about during human evolution has no impact whatsoever on my belief that we all bear God's Image, are morally responsible, and that we all are sinners in need of a Savior.

Evolutionary creationists believe that God's main purpose in the creation accounts of the Bible is to reveal inerrant spiritual truths. First and foremost, the opening chapters of Genesis teach us: (1) God created the world, (2) the creation is very good, (3) only humans were created in the Image of God, (4) all humans are sinners, and (5) God judges humans for their sins. These life-changing messages of faith are a gift from the Lord.

In contrast to young earth creation and progressive creation, evolutionary creation rejects scientific concordism. These Christian evolutionists believe that God accommodated and allowed the biblical writers to use the science-of-the-day as a vehicle to transport spiritual truths. As we saw in the last chapter, the Bible has ancient views of geography, astronomy, and biology (see Figures 5–3 and 5–9). In light of these ancient features, evolutionary creationists contend that the *de novo* creation of a 3-tier universe and the *de novo* creation of living organisms "according to their kinds" are ancient understandings of origins. Therefore, the creation accounts in Scripture are not a record of historical events revealing how God actually created the world.

Before we go further, I need to make a crucial comment about historical events that are recorded in the Bible. Though evolutionary creationists contend that the biblical accounts of origins are based on ancient scientific concepts, they firmly believe that real history in Scripture begins roughly around Genesis 12 with Abraham.

So from my perspective, Abraham was a real person. There really was a King David in Jerusalem during the tenth century BC. The Jews were deported to Babylon in the sixth century BC. And there definitely was a man named Jesus in the first century AD. The Gospels are eyewitness accounts of actual historical events, including the Lord's teaching and miracles, and *especially his physical resurrection from the dead*. Even though I do not believe the biblical creation accounts are a revelation of how the Creator made the universe and life, I thoroughly believe that the Bible records real events about the history of Jesus and his life.[10]

In addition to ancient science, evolutionary creationists note that God allowed ancient poetry in Scripture, such as the parallel panels of Genesis 1 (see Figure 2–2). Most would agree that real events in the past do not follow this poetic structure. In fact, the creation of light on the first day of creation before the sun on the fourth day is clear evidence of poetic freedom/license by the inspired author. He would certainly have known that light comes from the sun. But the parallel panels in Genesis 1 distinguish God forming boundaries in the universe during the first three days from his filling the world with heavenly bodies and animals in the last three days. Therefore, the writer had to place the creation of the sun in the second panel across from the origin of light in the first panel.

The most compelling feature of evolutionary creation is that it completely embraces *both* Christian faith and modern science. It meets the needs of our scientific generation in search of spiritual meaning. This view of origins offers an intellectually satisfying approach for those who experience God in a personal relationship and who know his creation through science. Evolutionary creation breaks us free from the chains of "either/or" thinking and the origins dichotomy. It frees us to be blessed by both the Book of God's Words and the Book of God's Works.

In understanding origins, evolutionary creationists embrace a complementary relationship between the Bible and science. Scripture and nature enrich and complete each other. The evolutionary sciences reveal *how* the Lord made this spectacular design-reflecting world. The Bible declares precisely *who* created it—the God of Christianity.

The greatest challenge faced by evolutionary creationists is that they do not accept the traditional and literal reading of the biblical creation accounts. Most Christians have understood these opening chapters of Scripture to be a historical record of God's actual miraculous events in creating the world. Evolutionary creationists suggest we need to move beyond scientific concordism. This takes time. Our churches and Sunday schools have entrenched in our minds the idea that there is an alignment between Scripture and science. With prayer and patience, I am convinced every Christian can be freed from concordism in order that they may focus on the inerrant spiritual truths in the Word of God.

The most important evolutionary creationist in the world is Dr. Francis Collins. He is the former director of the Human Genome Project that mapped out all the genes in the human body. Dr. Collins is not only a world-class scientist, but also one of the greatest scientists in the history of science. His bestselling book *The Language of God: A Scientist Presents Evidence for Belief*[11] is a fine introduction to evolutionary creation. In this book he also shares his spiritual voyage from atheism to faith in Jesus. Every Christian should be proud of Dr. Collins's incredible scientific accomplishments and in particular his amazing testimony of faith.

Deistic Evolution

Deism is the belief in an impersonal god. This divine being is often called the "god-of-the-philosophers." The deistic god is not involved in the lives of men and women. He never reveals himself personally through Scriptures, prayers, or miracles. In fact, the god of deism is a god that just doesn't seem to care about us. Regrettably, deistic evolution is sometimes mistermed as "theistic evolution." But properly defined, the word "theism" refers to belief in a personal God.

Deistic evolutionists claim that their god started the process of evolution with the Big Bang about 13.8 billion years ago, and then he stepped away from the universe never to return. This view of origins accepts cosmological evolution, geological evolution, and biological evolution. Deists picture their creator as one who winds up the world like a clock and lets it run down on its own without ever entering it.

By keeping their god outside the universe, deistic evolutionists firmly reject the personal God of Christianity. In particular, they completely refuse to believe that God became a man in the person of Jesus. They also deny that the Lord revealed inerrant spiritual truths through the inspired writers of the Bible. Deists claim that Scripture is basically a fairy tale that contains the fanciful religious ideas of ignorant ancient people. From this perspective, the Bible has no value whatsoever for our modern scientific generation.

The central problem with deistic evolution is that a god who winds up the clock of the universe and then walks away from it seldom meets the spiritual needs of anyone. The impersonal god of deism leaves most people cold, and rarely does this god move anyone to personal commitment and involvement. In contrast to Christianity, deism never inspired a lasting fellowship of believers, an educational institution, or a facility to help people in need, like a hospital or food bank. Deistic beliefs rarely transform people in the way the gospel of Jesus throughout history has led men and women to be born again and to dramatically change their lives.

Another issue with deistic evolution is that it seems quite odd that an impersonal god would create personal creatures such as us. We are incredibly relational. Just think about how many text messages people send to friends in just one day. Some of my students tell me they send out over 300 messages! It seems more reasonable to believe that the Creator is a personal God who is in a relationship with men and women. And this is exactly the God of Christianity. He reveals himself through the Bible, in answers to our prayers, and even with miracles.

The best example of a deistic evolutionist is the father of the theory of evolution Charles Darwin throughout most of his life. Today many Christians believe that he was an atheist. But this is incorrect. In his famous book *The Origin of Species*, Darwin refers seven times to a creator who made living organisms through evolution. As we will see in chapter 8, Darwin's actual religious beliefs are quite different from what is taught through secular education and in Sunday school.

Dysteleological Evolution

Dysteleological evolutionists claim that the universe and life evolved only through blind chance without any plan or purpose whatsoever. Regrettably, many people today believe that this atheistic interpretation of evolution is *the* evolutionist position and held by all scientists. Dysteleologists assert that humans are nothing but an unintended spin-off of biological evolution. In other words, our existence is just an accident and a mistake. These evolutionists believe that there is no ultimate right or wrong and that life is ultimately meaningless.

Atheistic evolution contends that anyone who believes in God is trapped in a delusion. God is merely a figment of human imagination. Dysteleological evolutionists often proclaim that they are the most logical thinkers in the world and that religious people are simply irrational and ignorant. They also believe that the Bible is just a fairy tale. Miracles like Jesus rising from the grave after his death are said to be nothing but fantasies concocted by wishful thinking. Atheists reject the spiritual truths in the Bible and view them as utter nonsense.

In my opinion, dysteleological evolution has a number of serious problems. Let me offer just a few reasons why I reject this view of origins.

First, atheism is not *the* official worldview of the scientific community. A well-known survey of leading scientists in the United States reveals that 40 percent of them believe in a God who answers prayer that is "more than the subjective psychological effect of prayer."[12] In other words, this is a personal God who intervenes in tangible ways in the lives of men and women. This survey also showed 40 percent of American scientists believe in life after death. It is not logical to claim that a significant percentage of the best scientists in the nation are irrational and ignorant.

Second, nearly everyone today rejects atheism. A survey of religious beliefs throughout the world reveals that only 1 percent of people are atheists.[13] In the United States, another study calculates that about 95 percent of men and women believe in the existence of God or a universal spirit.[14] This survey also discovered that only 1 percent of Americans accept atheism. Therefore, it is not reasonable for atheists to state that nearly everyone on earth today suffers from a delusion because we believe in God.

Third, dysteleological evolution guts out our human experience. Consider love. Atheists state that love is nothing but chemical activity in the brain. They assert that the idea of love is ultimately a delusion imposed on human relationships by misguided and soft-minded people. But does anyone who is in love believe that? Can you look at someone you love and say that the love you have for him or her is nothing but a delusion and that it doesn't really exist? Can you say that to your mom?

Finally, the greatest problem with dysteleological evolution is that it stands in the face of God and the first commandment, "You shall have no other gods before me" (Ex. 20:3; Deut. 5:7). This position commits the greatest of all sins. Atheists toss God away into the realm of delusion, and then they in effect put themselves in his place. They act like God and decide what is true, what is right, and what is wrong.

But in my opinion, taking the place of God is nauseating arrogance. Maybe atheists are trapped in the delusion that there is no God.

The most important dysteleological evolutionist in the world today is Richard Dawkins. One of his influential books is *The God Delusion*, and the title says it all. Dawkins believes that insulting Christians is productive. For example, in a television interview he referred to me as "an intellectual coward" and "a man with an air of desperation."[15] Needless to say, such an approach does not encourage a fruitful exchange of ideas. If anything, Dawkins has shown me that his reasons for being an atheist are for the most part emotional and have little rational value.

Summary: Relationships between the Positions on Origins

Figure 6–1 outlines the five basic categories on the origin of the universe and life that we discussed above. This chart demonstrates that there are more than just two simple positions on origins. It proves that the so-called "evolution" vs. "creation" debate is misguided and a mistake. This debate is a *false dichotomy*.

There are at least four different types of creationists and three kinds of evolutionists. Four of the positions believe that the world is the creation of a Creator—young earth creation, progressive creation, evolutionary creation, and deistic evolution. Three categories recognize that the inanimate universe and living organisms arose through evolution—evolutionary creation, deistic evolution, and dysteleological evolution.

Two views of origins believe in God and accept evolution—deistic evolution and evolutionary creation. However, these positions have radically different theological beliefs. Deistic evolutionists have an impersonal god and view Scripture as merely whimsical ideas of ignorant prescientific people. In contrast, evolutionary creationists embrace the personal God of Christianity and uphold that the Bible is the Holy Spirit-inspired Word of God.

	YOUNG EARTH CREATION Creation Science	PROGRESSIVE CREATION Day-Age Creation
Teleology	Yes	Yes
Intelligent Design	Yes Points to a Designer	Yes Points to a Designer
Age of the Universe	Young 6000 years	Old 13.8 billion years
Evolution of Life	No	No
God's Activity in the Origin of the Universe & Life	Yes Miraculous events over 6 days	Yes 1. Miraculous events for living organisms 2. Natural processes for inanimate universe
God's Activity in the Lives of Men & Women	Yes Personal God	Yes Personal God
Scientific Concordism	Yes	Yes
Interpretation of Genesis 1	Strict Literalism Creation days = 24 hrs	General Literalism Creation days = millions of yrs
Examples	Ken Ham Answers in Genesis	Hugh Ross Reasons to Believe

Figure 6–1. Positions on the Origin of the Universe and Life.

More specifically, the box outlined in Figure 6–1 identifies divine action from an evolutionary creationist perspective. These Christian evolutionists believe that God created the world through ordained and sustained evolutionary processes, and that he also acts personally in their lives with miracles and answered prayers. The line in the table reflects the Message-Incident Principle. Evolutionary creationists

EVOLUTIONARY CREATION Theistic Evolution	DEISTIC EVOLUTION "Theistic" Evolution	DYSTELEOLOGICAL EVOLUTION Atheistic Evolution
Yes	Yes	No Teleology an illusion
Yes Points to a Designer	Yes Points to a Designer	No Design an illusion
Old 13.8 billion years	Old 13.8 billion years	Old 13.8 billion years
Yes	Yes	Yes
Yes God uses ordained & sustained natural processes	Yes God uses natural processes	No Blind chance & natural processes God a delusion
Yes Personal God	No Impersonal God	No No God God a delusion
No	No	No
1. Spiritual truths 2. Ancient science 3. Ancient poetry	The Bible is a fairy tale	The Bible is a fairy tale
Francis Collins *Language of God* (2006)	Charles Darwin *Origin of Species* (1859)	Richard Dawkins *The God Delusion* (2006)

recognize that inerrant spiritual truths are delivered in Genesis 1 using the incidental vessels of ancient science and ancient poetry.

Figure 6–1 also reveals that there are three Christian positions on origins—young earth creation, progressive creation, and evolutionary creation. First and foremost, they are united by a belief that the world has an ultimate plan and purpose, and that this teleology originates from the Christian God. All three categories accept that nature features intelligent design and points to an Intelligent Designer. These Christian positions believe that God created the world and that he is personally involved in the lives of men and women. And they fully embrace the Bible as the Word of God. I think every Christian would say that these beliefs are essential to their faith.

The three Christian views of origins have different ideas on how the Creator made the world and on how we should read the creation accounts in Scripture. The key to understanding these differences rests with biblical interpretation. Young earth creationists and progressive creationists accept scientific concordism. They assume that there is an alignment between the Bible and the facts of science. In contrast, evolutionary creationists reject concordism. They claim that the Word of God has an ancient understanding of the natural world. As a result, it is impossible to align this ancient science in Scripture with our modern science.

From my perspective, knowing the method that God used to make the universe and life is not critical to being a Christian. Proof of this for me is the fact that I have met absolutely wonderful Christians who are young earth creationists, progressive creationists, and evolutionary creationists. They all love and serve the Lord with all their heart. I believe that our differences in understanding how God created the world should remain exactly that—different points of view. And these differences should never become a reason to divide us. We must always remember that we are united by Jesus and his love for us.

Fossil Pattern Predictions and the Christian Positions on Origins

One of the most exciting aspects of science for me is the testing of scientific theories to see if they are true. In this section, we will consider the basic fossil patterns predicted by the three Christian views of origins. Then the actual pattern of fossils found in the rock layers of the earth is presented to see if the predictions match up. It is important to point out that the sequence in which different fossilized plants and

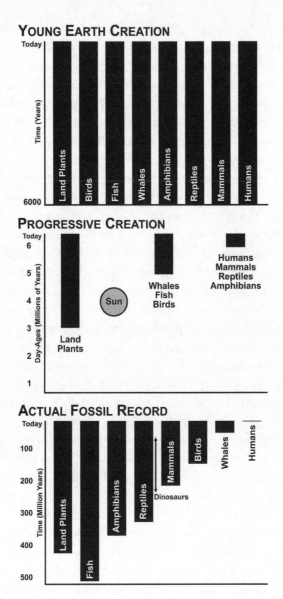

Figure 6–2. Fossil Pattern Predictions and the Fossil Record.
The arrow indicates where dinosaur fossils appear in the geological record.

animals appear in the geological record has been known for over 150 years, and rarely does anyone today doubt this pattern.

Figure 6–2 (top) presents the fossil pattern prediction of young earth creation. This view of origins claims that God created every type of living organism 6,000 years ago. Soon after the creation of humans, they sinned. In punishment for sin, God caused physical death to enter the entire world. Consequently, we should find at the bottom of the geological record a layer of fossilized bones with every kind of plant and animal that God made during the creation week. Notably, six day creation predicts that the bones of humans and dinosaurs (which are reptiles) should appear together in this lowest rock layer.

In addition, the Bible makes no mention of the extinction of any living organisms. Following the flood, God tells Noah to bring out all the animals from the ark "so they can multiply on the earth and be fruitful and increase in number" (Gen. 8:17). Clearly, it was never God's intention for any animals to go extinct, like the dinosaurs. According to a literal reading of Scripture, dinosaurs should be living with us today. In fact, why would God save them in the ark and then let them go extinct after the flood? Therefore, young earth creation predicts the accumulation of fossilized bones of every kind of creature through the entire fossil record, as indicated by the bars in the diagram.

Figure 6–2 (middle) shows the fossil pattern prediction of progressive creation. This position states that the creation days in Genesis 1 are periods that are millions of years long. Day-age creationists believe that God intervened miraculously to make different types of living organisms at different times during the 4.5 billion-year history of the earth. In particular, they claim that Genesis 1 reveals the order in which plants and animals were created. For example, old earth creationists believe that birds appeared on the fifth day/age of creation, millions of years before land animals on the sixth day/age.

It is worth pointing out that Genesis 1 states land plants were created on the third day/age. But the sun was created on the fourth creation day/age millions of years later. Clearly, without sunlight, plants would not have survived.

Figure 6–2 (bottom) is the actual pattern of fossils in the rock layers of the earth. As mentioned earlier, this sequence of fossils has been well-known for over 150 years, and it is accepted by nearly every scientist who studies geology. I think you will agree that the fossil pattern predictions of both young earth creation and progressive creation do not align with the scientific facts. Note that humans and dinosaurs never lived on earth at the same time as predicted by six day creation. And birds are not found before land animals in the geological record as expected by day-age creation.

It is important to observe that the actual pattern of animal fossils that we see in the rock layers shows an evolutionary sequence. Fish appear first, then amphibians, followed by reptiles (from which birds evolved), next land mammals (which led to whales), and last were humans at the top of the fossil record.

I am sure you noticed that there is no fossil pattern prediction for evolutionary creation in Figure 6–2. This is because evolutionary creationists reject scientific concordism. They do not believe that God put modern scientific facts in the Bible so that we can make scientific predictions. To state this in another way, evolutionary creationists reject the notion that the Bible is a book of science. Instead, they believe Scripture is a book that reveals God is Creator and he loves each of us.

What about You?

In my opinion, the key to understanding the origins debate is biblical interpretation. We often hear in our churches that there is no scientific evidence for biological evolution. But this belief is based ultimately on the assumption held by most Christians that Scripture features scientific concordism. Two Christian views of origins reject biological evolution—young earth creation and progressive creation. The main reason for their anti-evolutionism is concordism. They assume the Bible offers factual statements about how God created plants and animals quickly and completely through miraculous interventions. To be more precise, these anti-evolutionists believe in the *de novo* (that is, quickly and fully developed) creation of living organisms.

However, is it possible that the *de novo* creation of living creatures in Scripture is no different than the *de novo* creation of the firmament? Is the *de novo* creation of plants and animals similar to the *de novo* creation of the sun, moon, and stars in the firmament? If we do not use the Bible to determine the structure and origin of the heavens, then I don't think we should use it to understand how God created living organisms, including humans. From my perspective, the *de novo* creation of different forms of life "according to their kinds" in Genesis 1 is an ancient understanding of origins based on ancient biology.

Moving away from scientific concordism is difficult. I really struggled with the idea that the Word of God had ancient science. But as we saw in the last chapter, Jesus himself used ancient understandings of nature in his teaching ministry. He accommodated by employing ancient geography (Sheba at the ends of the earth, Matt. 12:42), ancient astronomy (stars fall to earth from the firmament, Matt. 24:29), and ancient biology (mustard seed the smallest seed, Mark 4:30–32). In this way, the Lord descended to the level of his ancient listeners and used their science as an incidental vessel to deliver life-changing messages of faith.

Similarly in Matthew 19:4–5, Jesus accommodated by employing the ancient science of the *de novo* creation of "male and female" in Genesis 1:27 to emphasize the inerrant spiritual truth that God is the Creator of human beings. Since God created us, we are accountable to him regarding how we live our life. Jesus also used the message of faith in Genesis 2:24 that through marriage a man and a woman "become one flesh." Therefore, Matthew 19:4–5 is not revelation of scientific facts on how God actually made humans. In this passage Jesus was responding to frivolous excuses justifying divorce. By appealing to Genesis 1:27 and 2:24, he argued that divorce was never God's intention for a husband and wife. As Jesus commanded, "What God has joined together, let no one separate" (Mark 10:9). Indeed, this is a message that our generation needs to hear and obey.

For me, Jesus is the foundation to understanding the creation accounts in Scripture. First, the Lord is the Ultimate Act of

Accommodation. He is the Creator of this incredible world, and he descended to earth and became a man to show his love for us. Second, Jesus also accommodated in teaching the Word of God. He employed incidental ancient sciences to reveal inerrant spiritual truths. It is in the light of Jesus that I believe God came down to the level of the inspired writers and allowed their ancient understanding of origins to be used in the biblical accounts of creation—the *de novo* creation of the universe and life, including men and women.

By now I am sure that you have figured out my view of origins. Yes, I'm an evolutionary creationist. But more importantly, what is your position on origins? Do you have an opinion regarding scientific concordism? That is the most critical issue that you need to deal with. At the beginning of this chapter I mentioned that it is perfectly acceptable to select and combine features from the five categories of origins presented in Figure 6–1. If you do so, you will have created a sixth origins position. And you will have rejected the origins dichotomy and moved beyond the so-called "evolution" vs. "creation" debate.

CHAPTER 7

GALILEO AND GOD'S TWO BOOKS

A number of people today view Galileo as the father of modern science. Yet he is probably better known for his conflict with the Catholic Church. Following the publication in 1632 of his most important book, *Dialogue Concerning the Two Chief World Systems*, he was put on trial for heresy. Church leaders believed the earth was stationary and situated at the center of the entire universe. They also assumed the sun was in motion and circled the earth. Galileo argued for a sun-centered world with the earth rotating on its axis and revolving around the sun. The church forced him to reject his scientific views in public and put him under house arrest for the rest of his life.

To be sure, the Galileo affair has been an embarrassment for the Catholic Church. Thankfully Pope John Paul II has apologized.[1] However, the damage was done. This historical episode has become the primary symbol of the popular understanding of the relationship between science and religion, known as the "conflict" or "warfare model." It pictures a battle between ignorant religious leaders armed with only their Bible on one side, and a brilliant astronomer equipped with a telescope and scientific evidence on the other. If we understand the story of Galileo in this way, we can appreciate why it continues to fuel the science vs. religion dichotomy. But in this chapter I will show that the warfare interpretation of the Galileo affair is too simplistic and not historically accurate.

History teaches us valuable lessons, and the story of Galileo is no exception. I want you to be aware that I have an agenda in this chapter. As we examine this historical event, I will be asking the question, "Is the modern origins debate a recycling of the Galileo affair with only the science in question being different—evolutionary biology instead of astronomy?" Currently, many Christians do not accept evolution. Are these anti-evolutionists like the church leaders who judged Galileo and rejected the notion that the earth circled the sun? We will also explore Galileo's remarkable approach to the relationship between science and Scripture. The famed astronomer offers helpful insights that move us beyond scientific concordism and the evolution vs. creation dichotomy.

Before looking directly at Galileo's views, we need to outline the two main scientific theories in his day dealing with the structure and operation of the universe. He refers to them in the title of that famous 1632 book as the "Two Chief World Systems."

Geocentrism (Greek noun *gē* means "earth") was the older astronomical theory. It was held by religious leaders in Catholic and Protestant churches as well as by most scientists at that time. This theory claimed the earth is at the center of the entire universe as depicted in Figure 7–1. Surrounding the earth is a series of spheres, each with its own planet. The motion of the spheres made the planets move. The moon was also in a sphere. The final sphere was the firmament with the stars attached to it.

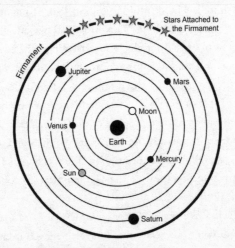

Figure 7–1. Geocentric Universe.

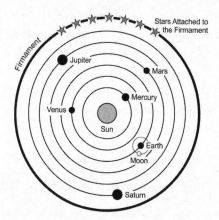

Figure 7–2. Heliocentric Universe.

In contrast, heliocentrism (Greek noun *hēlios* means "sun") places the sun in the middle of the world as shown in Figure 7–2. This newer theory was presented by Nicholas Copernicus in his 1543 book *On the Revolutions of the Heavenly Spheres*. As the title reveals, he also believed that spheres in heaven produced planetary and lunar motion. Galileo accepted Copernicus's heliocentric theory and popularized it in the early seventeenth century.

Two points need to be made about astronomy during this period. First, scientists did not have the concept of gravity to explain the movement of the planets and moon. It was not until the publication of Isaac Newton's *Mathematical Principles* in 1687 that they embraced the theory of gravity, leaving behind the notion that heavenly spheres caused the motion of astronomical bodies.

Second, a popular idea about the Galileo affair that continues to mislead many today is that the Catholic Church believed the earth was flat. Nothing could be further from the truth. In the early seventeenth century every astronomer accepted that the earth was a sphere, and there was no debate on that issue. The belief that Galileo battled church leaders who defended a flat earth is a myth concocted by anti-religious individuals in the nineteenth century as the warfare model of science and religion emerged.[2]

Now that we have some background information regarding the astronomy in Galileo's day, let's look directly at how he viewed the relationship between the Book of God's Words and the Book of God's Works. Galileo presents much of his approach to Scripture and science in his 1615 "Letter to the Grand Duchess Christina." I encourage every Christian to read this document. You will be amazed and blessed by it.[3]

God's Two Books: Foundational Principles

The Two Books Come from God

Galileo fully embraced the Book of Scripture and the Book of Nature. He believed the God of Christianity was the Author of the Bible and the Creator of the world. In his letter to the Grand Duchess, he writes, "For the Holy Scripture and nature derive equally from the Godhead, the former as the dictation of the Holy Spirit and the latter as the most obedient executrix of God's orders."[4] The term "Godhead" refers to the Holy Trinity—the Father, the Son, and the Holy Spirit. Galileo did not accept some sort of nebulous divine being, but specifically the God of Christian faith.

Notably, Galileo believed the Bible is the Word of God because it was inspired by the Holy Spirit. He also accepted that the laws of nature were ordained by God. According to Galileo, this world is a teleological world with an ultimate plan and purpose. If he were alive now, he would be appalled by the dysteleological metaphysics of Richard Dawkins and its conflation with modern science. Finally, in this passage Galileo presents a beautiful balance between God's Two Books. One is not greater or more important than the other.

The Two Books Reveal God

Galileo believed that Scripture and nature are divine revelations. He asserts, "God reveals himself to us no less excellently in [1] the effects of nature than in [2] the sacred words of Scripture, as Tertullian perhaps meant when he said, 'We postulate that God ought first to be known

[1] by nature, and afterward further known [2] by doctrine—[1] by nature through his works, [2] by doctrine through official teaching.'"[5] Tertullian was an important theologian who lived around 200 AD. Again, we see Galileo's marvelous balance between the Bible and the physical world. They are equally "excellent" in revealing our Creator.

Galileo had a strong view of natural revelation. He contends that those seeking to meet God should first open the Book of Nature and then afterwards examine "doctrine" and "official teaching" based on the Book of Scripture. To the surprise of many skeptics of religion, the father of modern science definitely believed in intelligent design. He acknowledged that the "effects of nature" point men and women toward a Creator and the "sacred words of Scripture" reveal this divine being is the God of Christianity. Clearly, Galileo embraced a complementary relationship between God's Two Books.

The Two Books Have Different Roles

Galileo recognized that the Book of God's Words and the Book of God's Works were different in that each had a specific purpose. As a consequence, they have their respective limitations. Taken together, these two divine books enrich and enhance each other in offering us an integrated revelation of the Creator, his creation, and us.

Galileo upholds the supremacy of the Bible over science when dealing with theological issues. "I have no doubt at all that, where human reason cannot reach, and where consequently one cannot have a science, but only opinion and faith, it is appropriate piously to conform absolutely to the literal meaning of Scripture."[6] This passage echoes the Metaphysics-Physics Principle in that religious belief is beyond the reach of science. Scientific investigations cannot extend into the spiritual realm. According to Galileo, the Bible offers spiritual truths. Faithful Christians should read these literally and embrace them fully.

With regard to matters dealing with science and the physical world, Galileo defends the priority of nature over Scripture. He writes, "I think that in disputes about natural phenomena one must begin not with the authority of scriptural passages but with sensory experience and necessary demonstrations [i.e., science]."[7] It is evident from this quotation that Galileo rejects scientific concordism and the notion that the Bible is a book of science. In order to understand the natural world, we need to open the Book of Nature, not the Book of Scripture.

Let's examine in more detail Galileo's beliefs regarding each of God's Two Books. First, we will consider his views about the physical world, and then his remarkable approach to interpreting the Bible.

Nature: The Book of God's Works

God Created Faithful Laws of Nature

Galileo believed natural laws were made by the Creator, and they have always operated according to his will. The famed astronomer asserts, "Nature is inexorable and immutable, [and] never violates the terms of the laws imposed on her. . . . [Nature is] the most obedient executrix of God's orders."[8] There are a number of subtle theological and scientific insights in this short passage.

First, the laws of nature function faithfully, and by implication they point to a Creator who is also faithful. In other words, God is not deceptive and we can trust the natural processes he created. For example, we get on airplanes because we believe natural laws are "inexorable and immutable."

Second, Galileo's belief in a God who created faithful laws in nature establishes a foundation for the practice of science. When scientists enter their laboratories, they trust that physical processes are consistent and repeatable. It would be impossible to understand how the world operates otherwise. By making a universe that, as Galileo writes, "never violates the terms of the laws imposed on her," the Creator has given us the opportunity to truly know his creation through scientific investigation.

Third, an implication of the physical world being run by "inexorable and immutable" natural laws is that God does not need to enter it and tinker around by adding missing parts or adjusting different processes. Nature is already a *fully gifted creation*! Consequently, the

Creator is not some sort of micromanager using sporadic God-of-the-gaps miracles to keep the universe and life functioning.

And to pique your interest, what other famous scientist embraced a view of the world similar to that of Galileo by claiming the laws of nature were "impressed on matter by the Creator"?[9] To the surprise of most people, this was Charles Darwin in his famous book on evolution, *The Origin of Species*. He also believed the laws of the physical world were God's laws.

Galileo's belief that natural laws were made by a faithful Creator inspires me. He allows us to consider that God "imposed" evolutionary laws on nature that she "never violates." Similarly, biological evolution can be seen as "the most obedient executrix of God's orders." And for me, the physical processes that led to the evolution of plants, animals, and humans offer evidence of the Lord's faithfulness.

Science Is a Gift from God

Galileo believed science ultimately comes from our Creator. He writes that when we possess information about nature "demonstrated with certainty or known by sensory experience [i.e., science] . . . it is a gift from God."[10] Of course, such a view of science clashes with that of atheists like Richard Dawkins. With nauseating pride, they see science as being entirely a human achievement. Instead of viewing it as a blessing from the Lord, atheists turn science into an idol they worship.

Galileo argued that the Creator gave us a mind so we could practice science. "I do not think one has to believe that the same God who has given us our senses, language, and intellect would want us to set aside the use of these. . . . Indeed, who wants the human mind put to death?"[11] Galileo affirms that God is not deceptive, but faithful. We can trust our mind and the scientific discoveries we make in nature because the Creator made us that way. An implication of being blessed by the Lord with "our senses, language, and intellect" is that he wants us to use these gifts. In fact, they assist us in obeying Jesus' commandment to love the Lord our God "with all our mind" (Matt. 22:37).

In the light of Galileo's belief that science and the human mind are gifts from God, we can view the study of evolution in a new way. From this perspective, evolutionary biology is a gift from the Lord. I doubt anyone wants to put their mind to death by not using their God-given intellectual abilities to understand how plants, animals, and humans evolved. And it is my personal experience that studying evolution is a way to love our Maker "with all our mind."

Nature Is Rational

Galileo claimed that God had inscribed rationality into the Book of Nature, and science was the instrument for reading this divine revelation. In *The Assayer*, he writes, "Philosophy [i.e., natural philosophy or science] is written in this *grand book*, the universe, which stands continually open to our gaze. But the *book* cannot be understood unless one first learns to comprehend the language and to read the letters in which it is composed."[12] Galileo then explains, "It is written in the language of mathematics, and its characters are triangles, circles, and other geometric figures without which it is humanly impossible to understand a single word of it; without these, one wanders about in a dark labyrinth."[13]

This remarkable passage leads to a number of questions regarding how God created life. What do we need to "first learn" before we can "comprehend the language" in the chapter on biological origins in the "grand book" of the universe? Perhaps geology and the fossil record? How about genetics and the DNA similarities between all living organisms? If we do not study these evolutionary sciences, will we "wander about in a dark labyrinth"?

Science Contributes to Biblical Interpretation

One of the most powerful insights that I learned from Galileo is that he believed the Book of Nature plays a role in understanding the Book of Scripture. Twice in his letter to Christina he states that scientific information assists in biblical interpretation. "Indeed, after becoming certain of some physical [scientific] conclusions, we should use these as very appropriate aids to the correct interpretation of Scripture."[14] Galileo adds that "it would be proper to ascertain the [scientific] facts first, so that they could guide us in finding the true meaning of Scripture."[15]

If we stop and think about it, every Christian today actually uses modern astronomy to interpret the Bible. For example, in verses referring to the immovability of the earth (1 Chr. 16:30; Ps. 93:1) and the daily movement of the sun across the sky (Eccl. 1:5; Ps. 19:6), I have yet to meet anyone who reads these passages literally as scientific facts. Most Christians interpret these verses as being phenomenological statements.

Why do they do this? It's because scientists have shown us the structure of the solar system and explained how gravity works. After the Galileo affair, Christians realized that astronomers had proven geocentricism to be false. The earth is not at the center of the universe with the sun circling it. Consequently, Christians could no longer read biblical verses about the earth's immovability and the sun's movement as factual scientific statements.

Let's apply Galileo's idea that science helps interpret Scripture to the issue of origins. Modern astronomy has contributed in recognizing the ancient astronomy in the Bible, and modern biology can assist us to see the ancient biology in Scripture. Specifically, the evolutionary sciences become "very appropriate aids to the correct interpretation" of the biblical creation accounts. In this way, evolutionary biology will help with the interpretation of Genesis 1 and 2 by identifying the ancient understanding of the origin of living organisms, including men and women.

Scripture: The Book of God's Words

Scripture Is Inerrant, Its Interpreters Are Not

Galileo embraced the notion of biblical inerrancy. In a letter to his student Benedetto Castelli in 1613, he proclaims that "Holy Scripture can never lie or err, and its declarations are absolutely and inviolably true."[16] But he qualifies, "Though the Scripture was inspired by the Holy Spirit . . . we cannot assert with certainty that all interpreters speak by divine inspiration."[17] Indeed! Galileo makes a critical point that every Christian needs to understand. The Bible is inerrant because it is inspired by God. The human interpretations of the Bible are not inerrant.

In the letter to the Grand Duchess Christina, Galileo further explains, "Holy Scripture can never lie, as long as its true meaning has been grasped."[18] An implication of his statement is that not all interpretations of the Bible are correct. Galileo argues that if every biblical interpreter was inspired by God, "then there would be no disagreement among them about the meaning of the same passages."[19] As everyone knows, there are lots of interpretive disagreements among Christians! The problem is not the Bible, but the interpreters of the Bible.

The fact that there exist in our churches a variety of interpretations regarding the biblical creation accounts is proof that Christians are not inerrant. Young earth creationists, progressive creationists, and evolutionary creationists present three entirely different ways to read Genesis 1. Simple logic dictates that at least two of these three interpretations are wrong. This fact should make all Christians pause for a moment, because *mistaken interpretations of Scripture are being taught in our churches and Sunday schools*. To correct this situation, I believe that learning the basic principles of biblical interpretation is a necessity for every preacher and teacher of the Word of God.

Scripture Deals with Salvation, Not Science

According to Galileo, the main purpose of the Bible is to reveal the path of salvation to men and women. "I should believe that the authority of Holy Writ has merely the aim of persuading men of those articles and propositions which are necessary for their salvation . . . [and] could not be discovered by scientific research or by any other means than through the mouth of the Holy Spirit himself."[20] Again we see Galileo's firm belief that God inspired the Bible. He also echoes the Metaphysics-Physics Principle in that the saving of souls is a spiritual issue that is beyond "scientific research." It is only through special revelation in Scripture that the Lord reveals directly the message of salvation.

Galileo also believed that God never intended to disclose scientific facts in the Bible. In his letter to Benedetto Castelli, he argues, "If the first sacred writers had been thinking of persuading the people about the arrangement and the movements of the heavenly bodies, they would not have treated of them so sparsely."[21] As we saw in chapter 5, if it had been God's intention to reveal scientific truths in Scripture, then it is reasonable to expect that he would have told us that our home is a spherical planet. The earth is mentioned about 2,750 times in the Bible, but never once is it described as a sphere or ball. Similarly for Galileo, the lack of astronomical information in Scripture is evidence that the Bible is not a book of science.

The message of salvation is quite simple, and it has nothing to do with science. As Acts 4:12 states, "Salvation is found in no one else, for there is no other name under heaven given to mankind by which we must be saved." Verse 10 identifies that the name of this person is "Jesus Christ of Nazareth." Romans 10:9 explains, "If you declare with your mouth, 'Jesus is Lord,' and believe in your heart that God raised him from the dead, you will be saved." Salvation deals with our relationship with Jesus, not with the age of the earth, not with the fossil record, and not with how God created the universe and life, including humans.

The Science in Scripture Is Ancient and Incidental

Even though Galileo firmly rejected the belief that the Bible was a source of scientific information, he recognized that there were "sciences discussed in Scripture" and noted that they were "the current opinion of those times."[22] He added that the Bible speaks "*incidentally*" of the earth, water, sun, or other created thing" since these are "not at all pertinent to the primary purpose of the Holy Writ, that is, to the worship of God and the salvation of souls."[23]

Clearly, Galileo accepted the basic concepts of the Message-Incident Principle. He refers to the ancient science in Scripture as "the current opinion of those times." Galileo also understood that this science-of-the-day is "incidental" and "not at all pertinent to the primary purpose" of the Bible. Instead, Scripture is intended to reveal inerrant messages of faith that will lead men and women to "the worship of God and the salvation of souls."

Galileo's view that the Bible has an incidental ancient science has significant implications for the origins debate. If Scripture has an ancient understanding of nature, then scientific concordism will never work. More specifically, if the biblical creation accounts feature an ancient conceptualization of origins, then Christian concordist positions like young earth creation and progressive creation are doomed to fail.

God Accommodated in Scripture

The Principle of Accommodation was a central idea in Galileo's approach to understanding biblical passages that refer to nature. He asserts that "propositions dictated by the Holy Spirit were expressed by the sacred writers in such a way as to *accommodate* the capacities of the very unrefined and undisciplined masses."[24] From Galileo's perspective, accommodation is part of the process that God used to inspire the Bible. It is worth noting that the term "accommodate" appears eight times in his letter to Christina. Galileo even observed that "this doctrine [of accommodation] is so commonplace and so definite among all theologians that it would be superfluous to present any testimony for it."[25]

Galileo contended that there was a good reason why the Bible does not include scientific facts. He argues "that it was necessary to attribute motion to the sun and rest to the earth in order not to confuse the meager understanding of the people and not to make them obstinately reluctant to give assent to the principal dogmas which are absolutely articles of faith."[26] If God had revealed in Scripture that the earth moved and the sun was stationary, *and God certainly could have done that*, it would have been a stumbling block (2 Cor. 6:3) to most ancient people. They would have had difficulty accepting "principle dogmas" of the Christian faith such as Jesus being fully man and fully God.

In light of Galileo's view on biblical accommodation, we can answer a question that Christians often ask today. Why does biological evolution not appear in the Bible? Such a revelation from God would have confused ancient people and been a stumbling block. Instead, the Lord graciously accommodated in Scripture by allowing the biblical writers to use their ancient understanding of origins—the *de novo* (quick and complete) creation of living organisms.

Galileo noted that the Principle of Accommodation was well-known in his day, and he thought that giving evidence for it in the letter to Christina was a waste of time. Regrettably, very few Christians today are aware of this interpretive principle. In my opinion, biblical accommodation is one of the most important ideas in developing a fruitful relationship between Christianity and science, and it needs to be taught in every Sunday school.

The Conflict between Galileo and His Enemies

As we have seen, Galileo enjoyed a peaceful relationship between his faith and science. He believed that scientific discoveries did not undermine the Bible because statements about nature in Scripture were accommodated and reflected the incidental ancient science of the biblical writers. However, there was conflict between Galileo and his enemies because they assumed his heliocentric astronomy attacked Christianity. Disagreements over biblical interpretation were the main cause of the Galileo affair. There were also other factors. As we examine some of these, let's keep in mind the modern origins debate and ask whether the opponents of the famed astronomer are similar to modern-day anti-evolutionists.

The Primary Problem: Scientific Concordism

Galileo's enemies were scientific concordists and used the Bible like a book of science. They assumed that statements in Scripture about the immovability of the earth and movement of the sun were scientific facts. Galileo exposes their concordism.

So the reason they advance to condemn the opinion of the earth's mobility and sun's stability is this: since in many places in Holy Scripture one reads that the sun moves and the earth stands still, and since the Scripture can never lie or err, it follows as a necessary consequence that the opinion of those who want to assert the sun to be motionless and the earth moving is erroneous and damnable.[27]

At a superficial level, there is a logical aspect to this argument. The Bible definitely states that the sun moves and the earth stands still. Christians believe that Scripture is inerrant. Therefore, it is reasonable to conclude that statements about the sun being stationary and the earth in motion are "erroneous and damnable." However, this argument is based on an ignorance of both Scripture and science. It is oblivious to the actual structure and operation of the universe, and it completely fails to respect that the Bible presents an ancient understanding of astronomy.

Do we have a similar problem today with anti-evolutionists and their scientific concordism? Let's recast Galileo's words within that context.

So the reason *anti-evolutionists* advance to condemn the opinion of the *evolution of life* is this: since in many places in Holy Scripture, *like Genesis 1 and 2*, one reads that *life was created de novo (quick and complete)*, and since the Scripture can never lie or err, it follows as a necessary consequence that the opinion of those who want to assert *the evolution of life* is erroneous and damnable.

Again, there is a superficial logic to this way of arguing. I can empathize with anti-evolutionists and their belief in concordism because more than thirty years ago, I often used this same line of reasoning. But I was quite ignorant of the basic evidence for evolution and very ignorant of the ancient biology in Scripture. My church and Sunday school had indoctrinated me with the assumption that scientific concordism was a feature of the Word of God.

Excessive Literalism and Proof-Text Interpretation

Galileo noted that reading the Bible literally all the time contributed to the conflict between him and his opponents. "Though the Scripture cannot err, nevertheless some of its interpreters and expositors may sometimes err in various ways. One of these would be very serious and very frequent, namely to want to limit oneself always to the literal meaning of the words."[28] As we noted in the last chapter, we have a similar situation today. Nearly 90 percent of born-again Christians in America believe the six day creation account in Genesis 1 is "literally true, meaning it happened that way word-for-word."[29]

Galileo also noticed that his enemies used proof-text interpretations. They ripped biblical verses out of their context and manipulated them to defend their anti-scientific views. Galileo charges, "They published some writings full of useless discussions and sprinkled with quotations from the Holy Scripture, taken from passages which they do not properly understand and which they inappropriately adduce."[30]

One of the best examples today of a proof-text interpretation is the misuse of the phrase "according to their kinds" in Genesis 1. Anti-evolutionists tear this phrase out of its ancient scientific context and claim it is biblical proof that God did not use evolution to create living organisms. But as we saw in chapter 5, "according to their kinds" is based on the ancient phenomenological perspective of plants and animals. It is an ancient taxonomical category that assumes creatures are immutable and do not change. This phrase is not a divine revelation against biological evolution—it's ancient science.

Misuse of Professional Authority

Another factor that fueled the Galileo affair occurred when well-educated enemies of the famed astronomer made claims outside their field of expertise. Galileo drives this point home with his metaphor of the absolute prince:

> Officials and experts of theology should not arrogate to them-
> selves the authority to issue decrees in the professions they

neither exercise nor study; for this would be the same as if an absolute prince, knowing he had unlimited power to issue orders and compel obedience, but being neither a physician nor an architect, wanted to direct medical treatment and the construction of buildings, resulting in serious danger to the life of the unfortunate sick and in the obvious collapse of structures.[31]

Professional theologians in Galileo's day should have known better. They were not astronomers and they were in no position to evaluate his astronomy. The ninth commandment states: "You shall not give false testimony against your neighbor" (Ex. 20:16; Deut. 5:20). To put it bluntly, Christians need to know what they are talking about before they speak. The theologians who put Galileo on trial for heresy and condemned him to house arrest for the rest of his life broke the ninth commandment.

The misuse of professional authority plagues the modern origins debate. Most of the leading anti-evolutionists do not even have biology degrees, let alone degrees in evolutionary biology. Try this as an exercise. Check out the education of people writing anti-evolutionary books. The majority have a doctoral degree in an academic field that is not related to evolution—law, engineering, history, philosophy, mathematics, astronomy, education, medicine, etc.

My suggestion to you is to always ask, "What type of doctor is the individual who is rejecting evolution?" I doubt anyone would seek medical treatment from a doctor who does not have a credible MD degree. So too in discussions about the origin of living organisms. Only let those with proper training in biology, especially evolutionary biology, be given the privilege of teaching about the origin of life in our churches and Sunday schools.

Disrespectful Attitudes

A lack of respect also contributed to conflict in the Galileo affair. For example, shameful attacks were launched against Nicholas Copernicus. Galileo observes, "Now that one is discovering how well founded upon clear observations and necessary demonstrations this doctrine is [heliocentrism], some persons come along who, without having even seen the book [*On the Revolutions of the Heavenly Spheres*], give its author [Copernicus] the reward of so much work by trying to have him declared a heretic."[32] I doubt that any Christian today would call another Christian a "heretic" for believing that the earth rotates on its axis and revolves around the sun.

There is a similar situation in our generation with Charles Darwin and his *Origin of Species*. Most Christians have not read the book, yet they feel quite confident in judging him to be the father of modern atheism. In fact, one of the most important anti-evolutionists in America once stated that "Satan himself is the originator of the concept of evolution."[33] But as we will see in the next chapter, if Christians would actually open Darwin's most famous book, they would see that he had a strong belief in both God and intelligent design.

Christians who attacked Copernicus and Galileo damaged the relationship between science and Christianity. It is my opinion that today young earth creationists and progressive creationists are harming an opportunity to share Jesus with evolutionists who are searching for the Lord. In addition, our churches and Sunday schools are discouraging young people from careers in the evolutionary sciences. In doing so, we are blocking their testimony of faith within the scientific community. We need to stop the disrespectful attitudes toward Darwin and biological evolution.

Galileo As a Symbol of Peace

The popular myth that anti-religious individuals continue to spew throughout our culture is that Galileo is the primary symbol of the warfare model of the relationship between science and religion. However, the actual historical evidence reveals how far this fictional story is from the truth. In reality, Galileo had a brilliant way of approaching the Book of God's Words and the Book of God's Works. There was never any hint that he felt his astronomy and faith were in conflict. The time has come for Christians to promote Galileo as a leading symbol of a peaceful relationship between science and Christianity.

From my perspective, the modern origins controversy is a recycling of the Galileo affair with the scientific issue being biological evolution instead of astronomy. The conflict in Galileo's day was not between the facts of science and the foundational beliefs of Christianity. It was caused mostly by the misguided assumption held by leaders of the Catholic Church that scientific concordism is a feature of Scripture. This problem also existed within Protestant churches at that time.

The central lesson we should draw from the Galileo affair is that every time the Bible is used as a book of science, the results will be disastrous for both modern science and our faith. If Christians today continue to read the biblical accounts of origins as scientific records of how God actually created the universe and life, we will only repeat the embarrassing mistakes of the church with Galileo. And worse, we will become stumbling blocks to those who know the evolutionary evidence and are seeking the Lord (2 Cor. 6:3).

The key to Galileo's peaceful relationship between science and faith is found in his remarkable approach to biblical interpretation. Echoing the Message-Incident Principle, he understood that the main goal of the Bible is to reveal inerrant spiritual truths regarding "the worship of God and the salvation of souls."[34] He also recognized that Scripture spoke "incidentally" about "created things" since these were "not at all pertinent to the primary purpose of the Holy Writ."[35] Galileo came to the conclusion that "the motion or rest of the earth or the sun are not articles of faith."[36] In applying these insights to the issue of origins, God's method of creating the world in Scripture is incidental, and evolution is not against the articles of our Christian faith.

Galileo's views on Scripture can be summarized by an aphorism he borrowed from Cardinal Baronio: "The intention of the Holy Spirit is to teach us how one goes to heaven, and not how heaven goes."[37] We can restate this saying within the context of the modern origins debate and suggest to our twenty-first-century generation:

The intention of Scripture is to teach us
who the Creator is, and not how he created.

CHAPTER 8

THE RELIGIOUS EVOLUTION OF DARWIN

To the surprise of most people, Charles Darwin offers us a great theological story. Throughout his life, he thought seriously about God and often wrestled with the religious implications of his theory of biological evolution. It may also be surprising to many that an essential aspect of the human spiritual voyage actually involves struggling with our Creator. In fact, the word "Israel" is made up of the Hebrew verb *śārâ*, which means "to struggle," and *'ēl*, a noun for "God."

The first time the name "Israel" appears in the Bible is when Jacob wrestled with God through the night. At daybreak the Lord said to him, "Your name will no longer be Jacob, but Israel, because you have struggled with God and with humans and have overcome" (Gen. 32:28). In my course on science and religion, many of my Christian students wrestle with the idea that living organisms have evolved. After learning about Darwin's theological struggles, several tell me that their faith was strengthened, and they can now consider biological evolution as the Lord's creative method.

Another surprising aspect of Darwin's story is that immediately following the publication of *The Origin of Species* in 1859, he grappled intensely with the notion of intelligent design in nature. By examining his struggles with this issue, we have the opportunity to test the view of design presented in chapter 4. I will interpret Darwin's many statements on design in light of the belief that biology declares the glory of God (Ps. 19:1). I will term this idea the "Psalm 19 Factor." Stated precisely, I will propose that Darwin's personal experience with living organisms gives us powerful evidence that intelligent design is real, and not merely an illusion as proclaimed by atheist Richard Dawkins.

The final surprise in this chapter will be to show that Charles Darwin did not embrace Darwinism! As we have seen, Dawkins equates the term "Darwinism" with his own dysteleological view of evolution in which "there is, at bottom, no design, no purpose, no evil and no good, nothing but blind, pitiless indifference."[1] In his bestselling book *The Blind Watchmaker*, Dawkins famously stated that "Darwin made it possible to be an intellectually fulfilled atheist."[2]

To counter Dawkins, I propose that Darwin makes it possible for me to be an intellectually fulfilled *theist*. Be assured that this is no attempt to "Christianize" Darwin, because he rejected Christianity as a young adult. But what I will suggest is that Darwin provides numerous theological insights that have assisted me as a Christian to embrace evolution as the Lord's creative process for making all plants and animals, including men and women.

The Early Years (1809–1831)

Charles Darwin was born 12 February 1809 in Shrewsbury, England. He grew up surrounded by a wide variety of religious, philosophical, and scientific ideas. His father, Robert, and older brother, Erasmus, were skeptics of religion.[3] Darwin's mother, Susannah, came from a Unitarian family and attended religious services with her children. Unitarians do not believe that Jesus was God, but only an enlightened man. Susannah died when Charles was only eight years old. Darwin's older sisters then brought him to an Anglican church where he was introduced to traditional Christianity. Charles was educated at an Anglican day school. As a teenager, he read his grandfather Erasmus Darwin's book *Zoonomia* (1794–1796), which presented a deistic god that created life through an evolutionary process.[4]

In 1828 Darwin entered Cambridge University to study theology. He recalls, "I did not then in the least doubt the strict and literal truth of every word in the Bible."[5] At that time in England, most Christians were anti-evolutionists because they read Scripture literally and accepted some form of scientific concordism. In particular, Cambridge introduced Darwin to a theological interpretation of the physical world. William Paley's *Natural Theology: Evidence for the Existence and Attributes of Deity, Collected from Appearances of Nature* was read in British universities. As the subtitle reveals, this book promoted the belief that the physical world was revelation from God. It was not long before Darwin was steeped in Paley's views of nature.

Darwin completed his theology degree in 1831, but decided not to be ordained as a pastor. Nevertheless, his Cambridge education was fruitful in that it gave him a purpose in life. During a geological field trip in Northern Wales, Darwin fell in love with science. There he experienced the power of scientific theories and their confirmation through discovering evidence in nature. Darwin's views on origins were typical of scientists in the early nineteenth century. He accepted that the earth was old and believed that God intervened to create life at different points through geological history. In other words, Darwin was a progressive creationist.

HMS *Beagle* Voyage (1831–1836)

After his Cambridge studies, Darwin boarded HMS *Beagle* in December 1831 and began a five-year trip around the world. Figure 8–1 maps out the voyage. In his 1876 *Autobiography*, Darwin comments that his Christian faith was "quite orthodox" at that time and recalls being mocked by the officers on the ship for quoting the Bible regarding moral issues.[6] It is during this expedition that he collected scientific evidence that eventually led him to formulate his theory of biological evolution.

When Darwin left England, he had a copy of the first volume of Charles Lyell's *Principles of Geology*. Lyell is often considered the father of modern geology. Not long after landing in Brazil, Darwin's research of the region convinced him of the superiority of this new approach to understanding the history of the earth. Lyell argued that geological evolution could be explained entirely through natural processes. There was no need for any miraculous God-of-the-gaps events

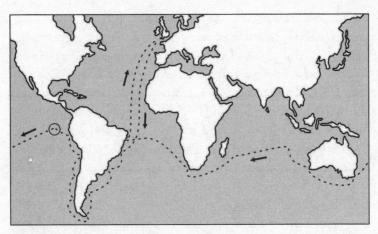

Figure 8–1. Darwin's Voyage on HMS *Beagle*. The circle indicates the region of the Galapagos Islands where Darwin discovered his famous finches.

like intermittent catastrophic floods, as some geologists at that time proposed. But did Darwin's natural-laws-only approach to geology extend to his biology?

No. Darwin continued to be a progressive creationist. In a revealing diary entry late in the voyage, he reflects upon the pitfall of an insect called the "ant lion" that he observed in Australia. The pitfall is a trap that catches ants. It is a conical depression made in sand by the ant lion, which then buries itself at the bottom just below the surface. Unsuspecting ants tumble down the sides and meet their death as ferocious jaws thrust out of the sand to crush them.

In thinking about the theological implications of a similar ant lion pitfall in England, Darwin asks, "Would any two workmen ever hit on so beautiful, so simple, & yet so artificial a contrivance [the pitfall]?"[7] He then answers, "It cannot be thought so. The one hand has surely worked throughout the universe. A Geologist perhaps would suggest that the periods of Creation have been distinct & remote the one from the other; that the Creator rested in his labor."[8]

Clearly, God was front and center in Darwin's science during this time. He definitely believed in an interventionist Creator whose "hand" had miraculously entered the world to create the ant lion with its ability to make a pitfall in both England and Australia. Darwin also believed that God's creative action was intermittent since there were "periods of Creation" followed by periods when "the Creator rested in his labor." Despite Darwin's acceptance of modern geology in which the history of the earth could be explained entirely through natural processes, he remained a progressive creationist with a God-of-the-gaps view of biological origins.

Darwin's description of the ant lion pitfall as "so beautiful" and "so artificial a contrivance" alludes to his acceptance of intelligent design. A contrivance is by definition something that is planned and crafted by an agent. Darwin's belief in design is even more evident in the final diary entry of his voyage:

> Among the scenes which are deeply impressed on my mind, none exceed in sublimity the primeval forests, undefaced by the hand of man, whether those of Brazil where the powers of life are predominant, or those of Tierra del Fuego, where death & decay prevail. Both are temples filled with the varied productions of the God of Nature:—No one can stand unmoved in these solitudes, without feeling that there is more in man than the mere breath of his body.[9]

Let's interpret this fascinating passage in the light of our biblical design categories previously presented.[10] First, the creation is active and its impact is universal. As Darwin states, the scenes of primeval forests are "deeply impressed" on his mind and "no one can stand unmoved" by them.

Second, the non-verbal message in the creation is intelligible. Darwin's use of the word "feeling" indicates that nature offers a revelation that does not use words, much like music. Nevertheless, Darwin understands the divine message because it leads him to the belief "that there is more in man than the mere breath of his body." In other words, he definitely rejects the dysteleological notion that humans are nothing but physical beings. Indeed, we are something more and Darwin later identifies this additional element to be the "immortality of the soul."[11]

Finally, the creation reveals God. The "varied productions" in the "primeval forests" point to the "God of Nature." Or, as Darwin later recalls of his experience in Brazil, the natural world led him "to the firm conviction of the existence of God."[12] But of course, the questions must be asked: Was Darwin experiencing the reality of the Psalm 19 Factor and a real divine revelation in nature? Or was this merely an illusion because he was indoctrinated by Christians at Cambridge University to perceive design in the world? I'll let you decide.

First Period of Religious Reflection (1836–1839)

Upon his return to England in October 1836, Darwin entered a period that he describes as the most productive years of his life. It was during this time that he outlined his theory of evolution. Alongside his scientific work, Darwin admits, "I was led to think much about religion."[13] Biological evolution has significant theological implications, especially for Christianity. In this first period of intense religious reflection, Darwin rejected Scripture, divine interventions, and the Christian faith.

Regarding the Bible, Darwin records in his *Autobiography*, "I had gradually come by this time, to see that the Old Testament from its manifestly *false history of the world*, with the Tower of Babel [Gen. 11:1–9], the rainbow as a sign [Gen. 9:13–15], etc., etc., . . . was no more to be trusted than the sacred books of the Hindoos, or the beliefs of any barbarian."[14] This is a painful passage to read. It is indeed sad that Darwin rejected the Word of God, but more troubling is his reason. He is one of the most important intellectuals in all of history, but this passage exposes his simpleminded approach to Scripture.

Darwin makes the massive assumption that the early chapters of Genesis are a historical record of actual events in the past. He assumes biblical accounts of origins feature scientific concordism and that they should be read literally as a "history of the world." However, if God allowed an ancient understanding of origins to be used in the process of inspiring Scripture, then Darwin's assessment of God's Word is terribly wrong. I am sure that his problem is obvious to you. Darwin lacked the most basic principles of biblical interpretation for passages dealing with the natural world.

Darwin also rejected God's intervention in both the origin and the operation of the world as well as in the lives of men and women. He offers the following argument against miracles:

> By further reflecting that the clearest evidence would be requisite to make any sane man believe in the miracles by which Christianity is supported—that the more we know of the fixed laws of nature the more incredible do miracles become—that the men at that time were ignorant and credulous to a degree almost incomprehensible by us, . . . by such reflections as these, which I give not as having the least novelty or value, but as they influenced me, I gradually came to disbelieve in Christianity as a divine revelation.[15]

It is important to note that Darwin rejects Christian faith at this time, not God. As we will see, he believes in a Creator who used evolution to create life when he wrote *The Origin of Species* twenty years later. Darwin's honesty is appreciated in that he acknowledges his arguments against miraculous events are not original. In fact, they are quite shallow. His stating that advances in science make divine activity "incredible" and claiming that only "ignorant" people believe in miracles do not align with the facts. As we saw in chapter 6, 40 percent of the best scientists in America accept that God answers prayer in a way that is more than just subjective and psychological.[16] I wonder what Darwin would say if faced with this hard evidence.

The main problem with Darwin's argument against miracles is that he lacks the basic categories of divine action. He makes the common mistake of carelessly conflating (blending) God's activity in nature with God's personal acts in the lives of men and women. I emphasized this critical categorical distinction in the table on origins in Figure 6–1 on pages 126–27. We need to separate how the Creator is involved in origins from his personal activity with us, and not conflate these two types of divine action.

By not having proper categories of divine action, Darwin makes a second huge assumption. He assumes that if there are no divine interventions in the origin and operation of the physical world, then God does not act in miraculous ways with people. But this is not necessarily true. It is perfectly reasonable to reject God-of-the-gaps events in nature as science demonstrates and to accept personal divine action. This is my experience both in practicing science professionally and in walking with the Lord personally. I suspect this view is held by the 40 percent of American scientists who believe in a personal God who answers prayer.

The Origin of Species (1859)

The publication of Charles Darwin's famous *Origin of Species* was the culmination of more than twenty years of work on his theory of biological evolution. Many Christians and anti-religious people presuppose that this book presents an atheistic worldview. Regrettably, few have actually read it. If they had, they would have been surprised to discover that there are seven positive references to the "Creator."[17] In fact, Darwin notes in his *Autobiography* that his belief in both God and intelligent design "was strong in my mind about the time, as far as I can remember, when I wrote the *Origin of Species*."[18] Let's look at a few revealing passages.

Darwin believed the Creator used natural processes to create living organisms. In responding to leading scientists of the day and their acceptance of progressive creation, he argues:

> Authors of the highest eminence seem to be fully satisfied with the view that each species has been independently created. To my mind it accords better with what we know of the laws impressed on matter by the Creator, that the production and extinction of the past and present inhabitants of the world should be due to secondary causes like those determining the birth and death of the individual.[19]

Clearly, Darwin believed that biological evolution was teleological. For him, the laws of nature are God's laws. In an earlier version of his book, he explicitly states, "By nature, I mean the laws ordained by God to govern the Universe."[20] Darwin would have completely rejected the dysteleological interpretation of evolution preached by Richard Dawkins. I suspect that Darwin would be insulted by the manipulation of his name when atheists use the misguided dysteleological term "Darwinism." During the writing of *The Origin of Species*, Darwin was never a so-called "Darwinist"!

This passage from Darwin's most famous book is a fine example of the Embryology-Evolution Analogy that we discussed in previous chapters. He contends that "secondary causes like those determining the birth" of each person are similar to those that led to "the production" of all "the past and present inhabitants of the world." Every Christian believes the Lord created each of us through his natural embryological processes. No one thinks that arms or legs were attached to our developing body in the womb through dramatic God-of-the-gaps interventions. Therefore, it is perfectly reasonable to believe the Creator "impressed on matter" evolutionary laws that led to the creation of every organism that has ever lived on earth.

In the last sentence of *The Origin of Species*, Darwin gives praise: "There is grandeur in this view of life, with its several powers, having been originally breathed into a few forms or into one; and that, whilst this planet has gone on cycling according to the fixed law of gravity, from so simple a beginning endless forms most beautiful and most wonderful have been, and are being, evolved."[21] This undoubtedly is an allusion to Genesis 2:7 where God breathed into the man the breath of life. In the next five editions of his book from 1860 to 1872, Darwin is even more explicit and replaces "originally breathed" with "breathed by the Creator."

Darwin's assessment that there is "grandeur" in the evolutionary view of life, resulting in creatures that are "most beautiful and most wonderful," naturally leads to the question of whether biological evolution reflects intelligent design. Is this evidence affirming the reality of the Psalm 19 Factor in biology? Or was Darwin merely experiencing an illusion that living organisms are designed? Again, you decide.

— Second Period of Religious Reflection (1860–1861) —

Immediately after the release of *The Origin of Species* in November 1859, Darwin entered a second two-year period of intense theological reflection. The central issue was intelligent design. To me, this is evidence that design in nature is not an irrelevant and outdated topic, as declared by some anti-religious individuals today. If the father of evolution struggled with design, we need to take note of it. A number of letters written to leading scientists of the day between 1860 and 1861 reveal Darwin's struggle and confusion with intelligent design.

The best of these letters was to Asa Gray. He was a devout Christian and a Harvard botanist who promoted *The Origin of Species* in America. In a letter dated 22 May 1860, Darwin states twice in no uncertain terms with regard to his book, "I had no intention to write atheistically. . . . Certainly I agree with you that my views are not at all necessarily atheistical."[22] It is worth pointing out that Richard Dawkins is aware of this letter since he cites it in one of his books.[23] But he conveniently overlooks these two clear statements that Darwin never intended to present an atheistic view of evolution.

In this letter to Gray, Darwin admits he is troubled by the existence of suffering in nature. He is particularly disturbed by a wasp that lays eggs in a caterpillar, and as these develop, they kill the caterpillar.

> But I own I cannot see, as plainly as others do, and as I should wish to do, evidence of design and beneficence on all sides of us. There seems to me too much misery in the world. I cannot persuade myself that a beneficent and omnipotent God would have designedly created the Ichneumonidae [wasp] with the express intention of their feeding within the bodies of Caterpillars, or that a cat should play with mice. Not believing this, I see no necessity in the belief that the eye was expressly designed.[24]

However, Darwin's doubt in design immediately disappears in the very next sentence. "On the other hand, I cannot anyhow be contented to view this wonderful universe and especially the nature of man, and to conclude that everything is the result of brute force."[25] In other words, this world is definitely *not* the dysteleological world of Richard Dawkins. Darwin's personal experience of the "wonderful universe" impacts him forcefully, as expected if the Psalm 19 Factor is true.

As a matter of fact, Darwin, in this letter to Gray, offers two evolutionary interpretations of intelligent design. With the first he asserts, "I am inclined to look at everything as resulting from *designed* laws, with the details, whether good or bad, left to the working out of what we may call chance."[26] From this perspective, God ordained the evolutionary process to create living organisms, but allowed some freedom with the origin of minor features. Personally, I find this approach to evolution and design quite persuasive.

In the second interpretation of evolutionary design, Darwin proposes, "I can see no reason, why a man, or other animal, may not have been aboriginally produced by other laws; and that all these laws may have been expressly *designed* by an omniscient Creator, who foresaw every future event and consequence."[27] According to this view, an all-knowing God designed evolutionary processes and knew exactly which creatures would evolve. I also find this view insightful in that it affirms the Creator's lordship over evolution.

In light of these two design interpretations, it is obvious that Charles Darwin never embraced an "evolution" vs. "design" dichotomy. He believed that it was perfectly logical to regard biological evolution as an intelligently designed process. If Darwin were alive today, he would dismiss the anti-evolutionism of the leaders of Intelligent Design Theory as a misguided view of origins.

The Descent of Man (1871)

When Darwin first outlined his theory of evolution in the late 1830s, he was quick to accept that humans evolved. In one of his notebooks he records, "I believe truer to consider him [man] created from animals."[28]

Twenty years later, in *The Origin of Species,* Darwin made only one short comment on this topic. He contended that with his theory of evolution, "Light will be thrown on the origin of man and his history."[29] But it is in his second most famous book, *The Descent of Man,* that Darwin presents a comprehensive theory of human evolution.

Being respectful of the many Christians in England, Darwin offers assistance to those troubled by the notion that humans have evolved from prehuman creatures. As in *The Origin of Species,* he again employs the powerful Embryology-Evolution Analogy.

> I am aware that the conclusion arrived at in this work will be denounced by some as highly irreligious; but he who denounces them is bound to show why it is more irreligious to explain the origin of man as a distinct species by descent from some lower form, through the laws of variation and natural selection, than to explain the birth of the individual through the laws of ordinary reproduction. The birth both of the species and of the individual are equally parts of that grand sequence of events, which our minds refuse to accept as the result of blind chance.[30]

Once again we have a passage from one of Darwin's most important books that reveals his metaphysical assessment of the evolutionary process—evolution is *not* "the result of blind chance." This is definitely *not* so-called "Darwinism," and it is *not* the dysteleological interpretation of evolution promoted by atheists like Dawkins. In fact, one could argue that the "grand sequence" of natural mechanisms in both embryology and evolution are natural revelations from the Creator reflecting intelligent design. From this perspective, God created us in such a way that "our minds refuse" to believe that evolution is due to blind chance. In other words, the human brain was made for seeing evidence of the Lord's existence through manifestations of design in the evolutionary process.

Without a doubt, the evolution of humanity is the most challenging topic in origins for Christians. But I believe that part of our difficulty with this theory is due to the fact that our only understanding of human evolution is steeped in the atheistic beliefs of Dawkins and his disciples. Darwin offers us a breath of fresh air in emphasizing that our evolution is not dysteleological. We are not a mistake or fluke of nature! Charles Darwin inspires us to consider that the Lord created men and women through an ordained and design-reflecting evolutionary process.

Autobiography of Charles Darwin (1876)

Darwin presents his mature theological views in a twelve-page section of his *Autobiography* entitled "Religious Belief." He offers reasons both "for" and "against" the existence of God. Each of these is followed by a counterargument. This pattern of argumentation leads to a stalemate, with Darwin being uncertain on whether or not God exists. He concludes, "I for one must be content to remain an Agnostic."[31]

Darwin deals directly with the greatest challenge faced by those who believe in a personal God—the problem of suffering in the world.

> A being so powerful and so full of knowledge as a God who could create the universe, is to our finite minds omnipotent and omniscient, and it revolts our understanding to suppose that his benevolence is not unbounded, for what advantage can there be in the suffering of millions of lower animals throughout almost endless time? This very old argument from the existence of suffering against the existence of an intelligent first cause seems to me a strong one.[32]

Yet Darwin quickly counters this argument. "According to my judgment *happiness* decidedly prevails. . . . All sentient beings have been formed so as to enjoy, as a general rule, *happiness*. . . . The sum of such *pleasures* as these, which are habitual or frequently recurrent, give, as I can hardly doubt, to most sentient beings an excess of *happiness* over misery, although many occasionally suffer much."[33] What a remarkable and helpful insight from Darwin. He paints a picture of the world that is far from the bleak and sinister view of nature promoted throughout our culture by atheistic evolutionists like Dawkins.

Though suffering in nature exists, Darwin trumps it and concludes that overall "happiness decidedly prevails."

The *Autobiography* also deals with the topic of intelligent design in nature. Darwin presents two interpretations of design and offers counterarguments in an attempt to dismiss both.

The first interpretation could be termed an "emotional" or "psychological design argument." Darwin observes, "At the present day the most usual argument for the existence of an intelligent God is drawn from the deep inward conviction and *feelings* which are experienced by most persons."[34] He then adds, "Formerly I was led by *feelings* such as those referred to" and these led him

> to the firm conviction of the existence of God, and of the immortality of the soul. In my Journal I wrote that whilst standing in the midst of the grandeur of a Brazilian forest, "it is not possible to give an adequate idea of the higher feelings of wonder, astonishment, and devotion, which fill and elevate the mind." I well remember my conviction that there is more in man than the mere breath of his body.[35]

These references to Darwin's experience in nature are consistent with many of our biblical design categories. The creation is active. The "grandeur of a Brazilian forest" impacted Darwin and convicted him. His use of the term "feelings" implies that this natural revelation is non-verbal. The creation's message is universal and never-ending since it is "experienced by most persons." Nature offers an understandable divine revelation. It led Darwin to believe in the "existence of God" and the "immortality of the soul."

But following the argument pattern in "Religious Belief," Darwin refutes this view of intelligent design. "But now the grandest scenes would not cause any such convictions and feelings to rise in my mind. It may be truly said that I am like a man who has become colour-blind."[36] Can anyone actually become color-blind to reflections of design in nature and not see God's glory? As we shall see, Darwin will provide a radically different answer in the last year of his life.

The *Autobiography* includes a second interpretation of intelligent design. Darwin finds this approach more convincing, and it could be termed a "rational design argument."

> Another source of conviction in the existence of God, con-nected with *the reason* and not with the feelings, impresses me as having much more weight. This follows from the extreme difficulty or rather impossibility of conceiving this immense and wondrous universe, including man with his capacity of looking backwards and far into futurity, as a result of blind chance or necessity. When thus reflecting I *feel* compelled to look to a First Cause having an intelligent mind in some degree analogous to that of man; and I *deserve* to be called a Theist.[37]

Though Darwin uses the word "Theist" in this passage, I believe the term "deist" would be more accurate. There is no evidence that he accepted a personal God at this point in time. Nevertheless, it is important to recognize that here in 1876, he uses the present tense for the verbs "feel" and "deserve." In other words, quite late in life, Darwin had periods when he believed in God. The reason for his belief was design in nature.

This passage also aligns well with our biblical design categories. First, the creation is active and impacts humans. The "wondrous uni-verse" not only "compelled" Darwin "to look" for a "First Cause," but he thought it was an "impossibility" to view the world "as a result of blind chance or necessity." Second, the message in the creation is intelligible. It reflects "an intelligent mind." Third, the creation reveals God. Nature points to a Creator with a "mind in some degree analogous to that of man." Or to put this notion within a Christian context, since we have been created in the Image of God, we should expect to be blessed with a mind that is in some small way similar to that of our Maker.

It is necessary to point out that immediately following the pas-sage above, Darwin writes, "This conclusion was strong in my mind about the time, as far as I can remember, when I wrote the *Origin* of Species."[38] To the surprise of most, and I suspect Dawkins as well, Darwin claims that he believed in God and intelligent design when he wrote his most famous book! How often do we hear this in the secular media? Never. In my opinion, this certainly exposes the so-called "objectivity" and intellectual integrity of secularists.

However, typical of his pattern of arguing in "Religious Belief," Darwin refutes his rational argument for intelligent design. "But then arises the horrid doubt—can the mind of man, which has, as I fully believe, been developed from a mind as low as that possessed by the lowest animal, be trusted when it draws such grand conclusions?"[39] I am sure you have spotted the problem with his refutation. What has Darwin just done? He has made a "grand conclusion" about human knowledge and the competence of the human mind. And what has Darwin just used to make this argument? He *trusted* his mind developed from the lowest animal! In other words, he blatantly contradicts himself. To use the technical term, Darwin's counterargument suffers from self-referential incoherence.

It is my opinion that Darwin does not offer a convincing refutation of his rational argument for design. Therefore, the "impossibility of conceiving this immense and wondrous universe, including man with his capacity of looking backwards and far into futurity, as a result of blind chance or necessity" remains a solid reason for belief in "a First Cause having an intelligent mind in some degree analogous to that of man." Thanks to Charles Darwin, I am equipped with a stronger view of intelligent design.

The Final Years (1879–1882)

Darwin's religious views in the last years of his life are basically similar to those presented in his *Autobiography*. Though he would say that he embraced agnosticism most of the time, he did have what appear to be periods during which he believed in some sort of god. This divine being seems to be the impersonal god of deism. There was also some shifting away from Darwin's so-called "colour-blindness" regarding intelligent design.

In one of his most important letters, Darwin responds to questions about his religious beliefs asked by John Fordyce. Darwin opens this letter, dated 7 May 1879, with a sharp criticism, "It seems to me absurd to doubt that a man may be an ardent Theist and an evolutionist."[40] Wow! In just one short sentence, Darwin completely destroys the origins dichotomy and the misguided assumption that people are forced to choose between biological evolution and their belief in a personal God, like the God of Christianity.

What an encouraging insight for Christians who accept evolution. Once again we have a startling statement from the father of evolutionary biology himself that few today would imagine he had made. Most people are not aware of Darwin's actual beliefs because anti-religious individuals like Richard Dawkins have manipulated the story of Charles Darwin for their atheistic agenda and misled the public away from the facts of history.

And there is more in this letter to set the record straight. Darwin then asserts, "In my most extreme fluctuations I have never been an Atheist in the sense of denying the existence of a God."[41] So there we have it, very late in life (he died in 1882) and well after the publication of his two most famous books, *The Origin of Species* and *The Descent of Man*, Darwin states in no uncertain terms that he was *never* an atheist and even defines the meaning of that term for us. One cannot help but wonder whether Dawkins and his so-called "Darwinist" disciples have ever examined the historical evidence regarding Darwin's actual beliefs.

Darwin adds, "I think that generally (and more and more so as I grow older) *but not always*, that an agnostic would be the most correct description of my state of mind."[42] Let's analyse this sentence and the previous one.

- Darwin was never at any time an atheist.
- Darwin was most of the time an agnostic.
- Therefore, during Darwin's "not always" periods, he was either a theist or a deist.

I suspect that Darwin was most likely a deist, since there is little evidence to suggest that he had a personal relationship with the Lord. The question might be asked, "What was the cause for these moments when Darwin believed in a divine being?" The next passage gives us the answer.

In the final year of Darwin's life, the Duke of Argyll engaged him on the issue of intelligent design in nature. In defending design, the Duke pointed to some of Darwin's extraordinary biological research:

> In the course of that conversation I said to Dr. Darwin, with reference to some of his own remarkable works on the 'Fertilization of Orchids' and upon 'The Earthworms,' and various other observations he made of the wonderful contrivances for certain purposes in nature—I said it was impossible to look at these without seeing that they were the effect and the expression of mind. I shall never forget Mr. Darwin's answer. He looked at me very hard and said, "Well, that often comes over me with *overwhelming force*; but at other times," and he shook his head vaguely, adding, "it seems to go away."[43]

From this passage it is obvious that Darwin miswrote five years earlier in his *Autobiography* when he claimed to be "colour-blind" to intelligent design in nature.

The conversation with the Duke of Argyll presents another opportunity to use our biblical design categories. The creation's message is active. As Darwin states, design in nature "comes over me with overwhelming force." The revelation in nature is intelligible. Orchids and earthworms could be seen as "the effect and the expression of mind." The message inscribed in the creation is incessant. Darwin confesses here late in life, it "often" came over him. The creation reveals the Creator. The "wonderful contrivances" and "purposes" found in living organisms point to a "mind" behind them. Finally, natural revelation can be rejected. Darwin sheepishly admits, design "seems to go away." But does it? Or is it because Darwin did not want to pursue the significant consequences of intelligent design?

For the last time in this chapter, let me ask you: Is the evidence from the writings of Charles Darwin affirming the reality of natural revelation and the Psalm 19 Factor? Or was Darwin simply experiencing an illusion of design in nature that was the result of his being indoctrinated by Christians in his generation? You tell me.

An Intellectually Fulfilled Theist

Richard Dawkins boldly claims that "Darwin made it possible to be an intellectually fulfilled atheist." However, the historical evidence presented in this chapter leads me to a completely different conclusion. Darwin's actual beliefs make it possible to be an intellectually fulfilled *theist*. In particular, the father of biological evolution offers many helpful theological insights. Here are some that have strengthened my worldview as an evolutionary scientist and Christian theologian.

First and foremost, evolution is not by necessity atheistic or dysteleological. In commenting on *The Origin of Species*, Darwin states that he "had no intention to write atheistically" and that his theory of evolution is "not at all necessarily atheistical." Darwin firmly rejected dysteleology and claimed the world is not "the result of blind chance or necessity." In fact, late in life he revealed that he had "never been an Atheist in the sense of denying the existence of a God."

Darwin's use of the Embryology-Evolution Analogy offers Christians a valuable way to understand God's creative action in biological evolution. This analogy appears in *The Origin of Species* and *The Descent of Man*. In the former, he views embryological and evolutionary processes as being the "laws impressed on matter by the Creator." In the latter, "The birth both of the species and of the individual are equally parts of that grand sequence of events, which our minds refuse to accept as the result of blind chance." Darwin leads us to consider that all natural processes, including the process of evolution, were ordained by God.

Another helpful Darwinian theological insight is that it is perfectly reasonable to accept biological evolution and intelligent design. Darwin admitted that belief in design was "strong in my mind" while writing *The Origin of Species*. He would undoubtedly dismiss the "evolution" vs. "design" debate promoted by the leaders of Intelligent Design Theory as a misguided dichotomy. Darwin also provides remarkable evidence for the reality of design, since he acknowledged the impact of nature throughout his life. Even in the year before his death, "the effect and the expression of mind" in biological structures struck him with "overwhelming force." As the Psalm 19 Factor states, biology declares the glory of God!

To conclude, the story of Charles Darwin is not only surprising, it is also an encouragement to Christians who are wrestling with biological evolution. He offers numerous theological insights that assist us to move beyond the "evolution" vs. "creation" debate. By declaring, "It seems to me absurd to doubt that a man may be an ardent Theist and an evolutionist," Darwin opens the door for us to love Jesus and to accept evolution. We can reword this remarkable insight for our twenty-first-century generation:

> *It seems absurd to doubt that anyone may be*
> *an ardent born-again Christian*
> *and an evolutionary creationist.*

CHAPTER 9

LET THE STUDENTS SPEAK!

One of the Lord's greatest blessings in my life is the privilege of teaching university students an introductory course on the relationship between science and religion.[1] As every teacher knows, our students teach us about as much as we teach them. Over the years I have learned valuable life lessons from my students. Many have faced challenges in their lives, and they have overcome these with courage and integrity. I am always humbled when they share their personal stories. My faith has definitely been strengthened by their spiritual voyages.

Most of the Christian students who enter my class say that to embrace their faith and accept modern science, they have to place each of these in a separate compartment. On Sunday mornings at church they are warned in sermons and Sunday school lessons of the dangers of the evolutionary sciences. But from Monday through Friday in their science classes, they are shown overwhelming evidence that the world is billions of years old and living organisms have evolved.

A number of my students are in pre-med, and they need to study genetics because this science is now becoming a central part of medicine. They soon learn that the genes in our body offer indisputable evidence for human evolution.[1] The only way they know how to deal with this situation is to compartmentalize these genetic facts from their religious belief that God created the first man and woman *de novo*. Yet intuitively they sense that there has to be a better solution, and they all yearn for an integration of Christian faith and modern science, including human evolutionary biology.

---

1   My entire course is online and accessible to the public for no charge. It includes 25 hours of audio-slide lectures, 200 pages of class notes, and 100 pages of class handouts. The course homepage is https://sites.www.ualberta.ca/~dlamoure/350homepage.html.

Of course, putting Christianity and science in separate compartments is ultimately due to my students being raised in churches and Christian schools entrenched in the science vs. religion and evolution vs. creation dichotomies. Early in my teaching career I discovered the importance of introducing students to a wide variety of approaches for moving beyond this "either/or" type of thinking. This has encouraged them to step away from compartmentalization and to construct a personal worldview that integrates their religious beliefs and scientific knowledge. Let me share a few stories about these remarkable students.

One of the most talented individuals ever to take my course was a brilliant young woman who should have been given an A+++ instead of an A+! As she walked out of my class on the last day, she said to me, "You taught me one thing." Of course, my pride was a bit wounded as I thought to myself, "Really, only one thing?" And then she said, "I am completely free. I have been freed from the dichotomies. And now I can love God and embrace evolution."

This was one of those thunderous moments in my teaching career that I will never forget. What more could I ask as a professor who both loves the Lord and believes he created through an evolutionary process? If any students who read this book walk away with this "one thing" of being "completely free" from choosing between science and religion, or between evolution and creation, then all the time and effort required to write it will have been worthwhile.

I have also seen the damage caused by the origins dichotomy. A geology student sat through my entire thirteen-week course and did not say one word. Afterward he came to my office and revealed that he entered my class with his "faith on a thread." On Sunday mornings he was being told the earth was only six thousand years old, but at university he was learning the scientific techniques to date rocks in the hundreds of millions of years. Most of his family were strong young earth creationists, and his growing acceptance of modern geology was creating serious tensions in his relationships with them.

The only way this geology student could make sense of life was to place his faith and science in separate compartments. The problem with compartmentalization is that it doesn't last long. Science usually displaces belief in God. This was exactly what was happening to this student prior to my course. The thin thread holding his faith was ready to snap. However, his dwindling belief in God stopped once he was introduced to various approaches of a peaceful relationship between Christianity and modern science. Like most Christian students, he was not aware of the scholarship in the new academic discipline of science and religion.[2]

But there is more to this story. I was invited to speak at a Bible school, and I asked him to come along and give a lecture on the age of the earth. I also wanted him to share how the origins dichotomy had created issues in his personal life. He agreed, and as we drove to the school I suggested that he could open his presentation in prayer so the students could know he was a Christian. I'll never forget what he said. "I'm not ready to pray yet." A knife cut through my heart. Damage caused by the origins dichotomy continued to linger in this student's soul. Sadly, most of it originated from his church and family.

Let me offer some more student stories. My course is similar to this book in that I believe a fruitful relationship between Christianity and science can be built on two basic concepts—the Metaphysics-Physics Principle and the Message-Incident Principle. Let's look at how each of these principles has impacted both Christian and non-Christian students.

The Book of God's Works and the Metaphysics-Physics Principle

I introduce the Metaphysics-Physics Principle on the second day of class. In doing so, I quickly challenge the common conflation of science and atheism and then start to undercut the science vs. religion

---

2   Two Christian science and religion organizations that offer students assistance in developing a healthy relationship between their faith and modern science include the American Scientific Affiliation (www.asa3.org) and the BioLogos Foundation (www.biologos.org). The ASA offers students a free membership, free registration at its annual conference, and free electronic access to journal, newsletter, magazine, and other website resources.

and evolution vs. creation dichotomies. This is a very freeing concept for Christian students in science programs, especially biology majors.

Only a couple days after learning this principle, a student came up to me after class. Her face was flushed and her eyes were glassy. I sensed something significant had happened. She said, "You have no idea what the Metaphysics-Physics Principle has done for me." I was a bit concerned and not sure what she meant. Then she added, "I no longer feel guilty when I study biology." Wow!

I knew exactly what this student was saying. She was being told in her church not to believe anything the university was teaching about the origin of living organisms. Many Christian students like her are instructed to learn about evolution in order to pass exams, but never to accept that evolution is true. What a terribly dysfunctional way to study biology! Even if a student discovers scientific evidence that supports evolution, they have to deny what they see with their own eyes. Freed from conflations and dichotomies, my student now praises, "I look forward to biology class because it is really exciting to learn about how the Lord created through evolution." Double wow!

It's always amazing to me that about 10 percent of my students are atheists and agnostics. My class is listed as a theology course, but nevertheless they show up. I have found over the years that these religious skeptics always make valuable contributions in our class discussions. One student stands out in my mind. She raised her hand one day and said, "I've got something to say." I replied, "Please do." Then she said, "First of all, I want you to know that I'm an atheist, and I don't buy any of this Jesus stuff." I returned, "That's no problem. You've got total intellectual freedom in my classroom . . . and I'll protect you from the Christian horde!"

The student continued, "I also want you to know that my major is evolutionary biology. The Metaphysics-Physics Principle has opened my eyes to see that evolution is not necessarily atheistic. With this being the case, I am now totally appalled with the way many of my professors come into class and misuse their scientific authority and try to 'baptize' evolution with their atheism, giving the impression that atheistic evolution is *the* official view of science."

Needless to say, the Christians in the class were squirming in their seats and couldn't believe what they had just heard from an atheist. Being in a public university, religious students are often overwhelmed by secular and anti-religious ideologies that are often passed off as being "*the* Truth" and "*the* scholarly position." From my point of view, science classes should be teaching only science and not being used as a platform for some scientist to preach their naive metaphysical beliefs.

I have also noticed a pattern over the years with a number of my atheist and agnostic students. Usually a couple weeks after being introduced to the Metaphysics-Physics Principle, some make an appointment to visit me in my office. This is a special moment for me. It's a holy moment. These students confess that this principle has made them realize that they are no different than the religious students in the class because everyone, including themselves, has to take a step of faith from their science to their ultimate metaphysical beliefs.

I am always in awe when these students share this thought with me. Their courage and intellectual honesty is inspiring. In my opinion, coming to the understanding that atheism and agnosticism are beliefs, and not scientifically provable, is the most important idea that the non-religious students learn in my course. They often read in popular books and on the Internet that science proves there is no God. But science deals only with the physical, not the spiritual or metaphysical.

One last story about the Metaphysics-Physics Principle, and yes, it involves another of my remarkable atheistic students. He was the leader of class discussions and everyone loved him because he bubbled with joy and always had a smile on his face. I had numerous debates with him and always appreciated his respectful challenges to my faith. He loved biology and eventually became a professional biologist.

In one of the class discussions, he told us that in middle school he was continually teased and mocked by young earth creationist students. He said, "If I would have been allowed to believe that God created plants and animals through evolution, I probably would not have become an atheist." Ouch. Here was another heart-crushing moment in my teaching career. Can we learn from this story? Are Christian anti-evolutionists a stumbling block (2 Cor. 6:3) between God and evolutionists who are searching for him?

The Book of God's Words and the Message-Incident Principle

After every course at my university, students are given a questionnaire to evaluate their professor. One question asks to identify the most valuable aspect of the class. The majority of my students state that learning the concepts of biblical interpretation was the best part of the course. They often point to the Message-Incident Principle, the Principle of Accommodation, and the problem of scientific concordism as being very helpful concepts. For most students, the assumption that the Bible is supposed to align with the facts of science was a long-standing issue for them. Once freed from concordism, many experience a renewed love for Scripture and its inerrant spiritual truths.

Challenging the literalist and concordist interpretations we learned in church and Sunday school is not at all comfortable. Let me give you an example of a student in my very first evening class. This young woman was a delightful Christian and a staunch young earth creationist. She was also a freshman who did not realize that she could drop a course and transfer into another one. She assumed she was stuck in my class for the entire term.

After completing my course, she told me that in the first half of the term, she would walk home at night crying. Ugh! I felt awful. She was deeply troubled by the biblical evidence presented in class that challenged the literal reading of Genesis 1 she had learned in Sunday school. She also confessed that she had a mantra referring to me that she repeated on the way home: "I hate him! I hate him! I hate him!"

However, everything changed dramatically in the middle of the term once she was introduced to Galileo and his views on biblical interpretation in the "Letter to the Grand Duchess Christina." She realized that her issues with evolution and Scripture were no different than those regarding astronomy and the Bible in Galileo's day. In particular, she discovered the problem with scientific concordism. It is impossible to align the ancient astronomy in Scripture with modern astronomy. With this information, it was easy for her to see that the Bible has an ancient view of the origin of living organisms, and as a consequence, Scripture does not reveal how God actually created plants and animals.

I am very pleased to report that in the second half of my course, this young woman went home in the evening with a new mantra: "I love him! I love him! I love him!" I am assuming she was referring to Galileo!

Occasionally pastors from local churches take my science and religion course. Near the end of the term, I always ask them in front of the class, "What percentage of the biblical interpretation principles presented in my course did you learn in your three years of theology school?" The figure is always quite low, only around 25 percent. I use this moment to encourage the younger students to be patient with their church leaders, because some students get very angry with them, like the young woman I described on the first page of this book. Most pastors are not aware of non-concordist approaches to passages in the Bible that refer to nature. It's going to take some time to correct this situation.

The Christian students tell me that the concepts for interpreting Scripture, such as the Message-Incident Principle and the Principle of Accommodation, are not difficult to understand and that these could easily be taught in Sunday school. I agree. But again, moving beyond scientific concordism and literalist interpretations of the opening chapters of the Bible will take time. I know this personally, because I struggled for many years with this issue. We all have to be patient. I am convinced that stepping away from concordism will take at least a generation, and I am certain that in the future, individuals like my students will lead the way with respect and integrity.

The importance of introducing principles of biblical interpretation in Sunday school can be seen in this next example. Toward the end of my course, a student wrote me a note after a class discussion on origins.

I think that the conflict between the Bible and modern science taught in Sunday school is part of the reason I lost my faith a long time ago. Maybe, if we had been taught principles of biblical interpretation in Sunday school, I would still have my faith today. As I am now, being raised in a mostly literal interpretation of the opening chapters of the Bible, I am too critical of Christian faith, and I don't think I will return.

My heart still aches when I read this note. This is real life. Spiritual tragedies like this are happening throughout the nation today. Christian leaders need to address this disastrous situation.

It is my experience that young people are very teachable. In my opinion, all that church leaders need to do is present the various options and not force upon them only one view, like the literalism and scientific concordism of young earth creation. Teach the different views in a fair and balanced way, and young men and women will figure it out on their own. I have a great trust in them, because in my class I see them making good decisions time and again.

By the end of my thirteen-week course, about 95 percent of my students come to accept both the Metaphysics-Physics Principle and the Message-Incident Principle. In many ways, my classroom is like a laboratory for testing new ideas. As every university professor knows, if we ever say something that is illogical or contradictory, students are very quick to challenge us. The fact that such a high percentage of both Christian and non-Christian students embrace these two foundational principles suggests to me that these concepts are true and resonate deeply with their intellectual and personal life experiences.

Let me close this series of student stories with the two basic responses that Christians express after taking my course on science and religion. The first is caution. About a third of these students tell me that everything they have learned "makes perfect sense." They can appreciate evolutionary creation as a Christian view of origins, and they fully understand the problem with scientific concordism. They now recognize that the Bible is not a book of science. However, they feel uncomfortable with these ideas and are not ready to accept them.

I am quick to remind these Christian students that there is no rush to embrace any of these concepts because everyone processes them at different speeds. But more importantly, I emphasize that Christianity is about a personal relationship with Jesus, and not about how the Lord created the world. It's okay to feel like these students feel. Be cautious. Take your time. There is no hurry.

The second response to my course comes from about two-thirds of the Christians. They are quite grateful after taking the class. Many science majors have been looking for a view of origins that integrates their faith with evolution. When introduced to evolutionary creation, they found that "all the pieces fell into place." A number of Christian students reveal the course has taught them the meaning of loving God with our mind, as Jesus commands in Matthew 22:37. That humbles me. Even more humbling are the students who say their faith has been strengthened in my class. Irony of ironies. An evolutionist professor strengthening the faith of Christian students! Who would have thought?

Evolution: Scripture and Nature Say YES!

I am certain that you have figured out the title of this book. It is only when Scripture and nature are taken *together* in a complementary relationship that they can say "yes" to evolution. In dealing with the origin of the universe and life, God's Two Books complete one another in that each adds something not found in the other. The Book of God's Words reveals spiritual truths. The Book of God's Works offers scientific facts. The Bible tells us *who* created and science shows *how* he created. Together these two divine books provide an integrated revelation of our Creator, his creation, and us.

Figure 9–1 depicts the reciprocal relationship between God's Two Books, with special attention to the evolutionary sciences in cosmology, geology, and biology. This diagram incorporates the two foundational concepts of this book—the Metaphysics-Physics Principle and the Message-Incident Principle. From my perspective, there is a two-way exchange of ideas between the Bible and evolution.

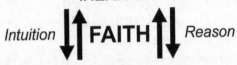

BOOK OF GOD'S WORDS
Spiritual Truths
INERRANT

Intuition ⬆⬇ FAITH ⬆⬇ *Reason*

BOOK OF GOD'S WORKS
Scientific Facts
Evolutionary Sciences
Cosmology, Geology & Biology

Figure 9–1. A Complementary Relationship between God's Two Books, with Emphasis on the Evolutionary Sciences.

Each informs and strengthens the other in understanding the origin of the world. Let me explain.

As the downward arrows in the diagram indicate, the Book of God's Words through a step of faith reveals inerrant spiritual truths about the universe and life. First and foremost, Scripture tells us that the Creator of this amazing world that scientists study daily is the God of Christianity. The Bible also reveals that the creation is very good, and by implication we need to take care of it. And the Word of God declares that the stunning beauty, intricate complexity, and astonishing functionality discovered through microscopes and telescopes are reflections of the intelligent design inscribed on nature by the Lord. In this way, the Book of Scripture sheds light on the physical world through the argument *from* design *to* nature.

The Bible also states that our Maker ordained the creation of the physical world, and that he has sustained it throughout time. By an act of faith, these spiritual truths lead us to identify the Christian God as the Ordainer and Sustainer of the process of evolution. Moreover, Scripture reveals the inerrant beliefs that humans bear the Image of God, are morally accountable, and have all sinned. This divine revelation leads evolutionary creationists to believe that these spiritual realities were manifested during human evolution with the appearance of the first men and women.

The upward arrows in Figure 9–1 indicate that through intuition and reason, and ultimately by a step of faith, the Book of God's Works offers scientific facts that inform and strengthen our belief in both God and Scripture. As the history of science reveals, every scientific discipline, including the evolutionary sciences, has presented countless facts of nature to support the biblical belief that the world is a natural revelation reflecting intelligent design. The incredible self-assembling character of cosmological, geological, and biological evolution cries out that an Intelligent Designer exists, and this evolutionary evidence confirms Scripture in that this Creator is glorious, divine, and eternally powerful (Ps. 19:1; Rom. 1:20). The Book of Nature bolsters our belief in the God of the Bible through the compelling argument *from* nature *to* design.

In addition, science today offers Christians an infinitely more magnificent picture of our Creator and his creation than ever before. We no longer think that we live in a diminutive 3-tier universe or a geocentric world enclosed by a spherical firmament. Instead, astronomy reveals an unbelievably enormous cosmos with about 100 billion galaxies and around 100 billion stars in each galaxy! In the future I fully expect science to continue this trend of revealing greater manifestations of God's glory and workmanship in nature. Scientific discoveries have also assisted in the interpretation of Scripture. It is only because of modern science that we now understand the Lord accommodated in the Bible by allowing the inspired writers to use ancient science, the best science-of-the-day thousands of years ago.

This is necessary to repeat: The key to a fruitful relationship between God's Two Books is that they must be taken *together*. Standing alone they offer only an incomplete revelation of our Creator, the creation, and humanity. For example, Christians limiting themselves just to Scripture will be forced to conclude that they live in a 3-tiered universe and that God created such a world *de novo*. On the other hand, if someone focuses merely on nature, it leads to an unknown and nebulous creator. It is only by embracing *both* the Book of Scripture and the Book of Nature that we can enjoy the fullness of the two divine revelations that God intended for us.

The epigraphs at the front of this book were written by Sir Francis Bacon, and they present a complementary relationship between God's Two Books. Bacon was both a devout Christian and one of the founding fathers of the scientific method in the seventeenth century. He believed that Scripture reveals the "will of God" and that nature displays the "almighty power" of the Christian Creator. In particular, Bacon recognized that studying the creation is "a key" to understanding the meaning of the Bible. He also accepted intelligent design since the "works" of God were "signed and engraved" by him. Bacon cautions us never to think that we "can search too far or be too well studied" in Scripture or nature. Instead, he encourages "everyone to endeavor an endless progress or proficiency in both."

Notably, Bacon appeals to Jesus' criticism of religious leaders in Matthew 22:29: "You are in error because you do not know the Scriptures or the power of God." Can the Lord's words be redirected at Christians today trapped in "either/or" thinking and entrenched the origins dichotomy? To scientific concordists, "You do not know the Scriptures" because God accommodated and allowed an ancient understanding of origins to be used in the biblical accounts of creation. Or to young earth creationists and progressive creationists, "You do not know the power of God." Our Creator has the unfathomable foresight and strength to create a self-assembling world, culminating with the evolution of men and women who bear the Image of God.

By now you know my position on concordism and anti-evolutionism. But what do you think? More importantly, how do you relate the Book of God's Words and the Book of God's Works? Does each inform and enhance the other? And one last question: If you hold Scripture and nature *together* in a complementary relationship, do they lead you to say "yes" to evolution? I look forward to your answer.

LIST OF DIAGRAMS

ENDNOTES

Epigraph

1. Francis Bacon, *Of the Advancement and Proficience of Learning* (Oxford, UK: University Press, 1640 [1605]), 9, 47. Passages enumerated as 1.1.3, 1.6.16. Language adapted from the original seventeenth-century English. Original text at https://ia600407.us.archive.org/17/items/ofadvancementp00baco/ofadvancementp00baco.pdf.

Chapter 1: Trapped in "Either/Or" Thinking

1. "Six Reasons Young Christians Leave Church" (September 28, 2011), www.barna.org/barna-update/millennials/528-six-reasons-young-christians-leave-church.
2. Information about the Institute for Creation Research can be found at www.icr.org.
3. Denis O. Lamoureux, "Philosophy vs. Science," *Creation Science Dialogue* 8:3 (Winter 1981): 3. Also at www.ualberta.ca/~dlamoure/p_yec.jgp.

Chapter 2: Opening God's Two Books

1. To emphasize the thrust of the original Hebrew language, I have replaced the term "vault" in the 2011 NIV with "firmament," as well as "sky" with "heaven." I have also pluralized "water" to reflect the Hebrew dual. Similar replacements occur elsewhere in this book.

2. In chapter 5 we will examine in more detail this original biblical term.

3. See Figures 5.6 and 5.7 on pages 102 and 103.

4. Adapted from "The Bible is the Word of God given in the words of men in history," in George Eldon Ladd, *The New Testament and Criticism* (Grand Rapids: Eerdmans, 1967), 12.

5. Duane Gish, *Evolution: The Fossils Say No!* (San Diego: Creation-Life, 1972), 33.

6. Robert L. Carroll, *Patterns and Processes of Vertebrate Evolution* (Cambridge, UK: University Press, 1998), 300. Redrawn by Andrea Dmytrash. Body outline of amphibian drawn by paleontologist Michael W. Caldwell. Top: *Acanthostega*; bottom: *Eusthenopteron*.

7. M. I. Coates, J. E. Jeffrey, and M. Rut, "Fins to Limbs: What the Fossils Say," *Evolution and Development* 4 (2002): 392; Edward B. Daeschler and Neil Shubin, "Fish with Fingers?" *Nature* 391 (January 8, 1997), 133; Carroll, *Patterns and Processes*, 233. Redrawn by Andrea Dmytrash. Left: *Eusthenopteron*; middle: *Sauripterus*; right: *Acanthostega*.

8. Drawn by paleontologist Braden Barr.

9. Robert L. Carroll, *Vertebrate Paleontology and Evolution* (New York: W. H. Freeman, 1988), 196, 365, 386, 406, 408. Redrawn by Braden Barr. Top to bottom: *Morganucodon, Dimetrodon, Cynognathus, Protorothyris*.

10. Top jaw from Kenneth D. Rose, *The Beginning of the Age of Mammals* (Baltimore: John Hopkins University Press, 2006), 92; bottom three jaws from Carroll, *Vertebrate Paleontology*, 366, 382, 390. Redrawn by Andrea Dmytrash. Top to bottom: *Daulestes, Probainognathus, Thrinaxodon, Haptodus*.

11. The four whales are from Rose, *Mammals*, 282; the hind limb is from Philip D. Gingerich, B. Holly, and Elwyn L. Simons, "Hind Limbs of Eocene *Basilosaurus*: Evidence of Feet in Whales," *Science* 249 (July 13, 1990), 155. Redrawn by Andrea

Dmytrash. Top to bottom: *Basilosaurus, Dorudon, Rodhocetus, Ambulocetus.*

12. Max Weber, *Die Säugetiere*, reprinted (Jena, Germany: Von Gustav Fischer, 1904), 558. Bowhead whale (*Mysticetus*).

13. A. E. W. Miles and Caroline Grigson, eds., *Colyer's Variations and Diseases of the Teeth of Animals* (Cambridge, UK: University Press, 1990), 101. Redrawn by Braden Barr. Left: rorqual whale (*Physalus*); center: blue whale (*Musculus*); right: minke whale (*Rostrata*).

Chapter 3: Terms That Begin to Free Us

1. Richard Dawkins, *River Out of Eden: A Darwinian View of Life* (New York: Basic Books, 1995), 133.

2. I have debated Johnson, Behe, and Meyer. See Phillip E. Johnson and Denis O. Lamoureux, *Darwinism Defeated? The Johnson-Lamoureux Debate on Biological Origins* (Vancouver, BC: Regent College Press, 1999); Michael J. Behe, "Design vs. Randomness in Evolution: Where Do the Data Point?" *Canadian Catholic Review* 17:3 (July 1999): 63–66; Denis O. Lamoureux, "A Black Box or a Black Hole? A Response to Michael J. Behe," *Canadian Catholic Review* 17:3 (July 1999): 67–73. My March 2016 debate with Stephen Meyer can be found at www.youtube/watch?v=Muy58DagOK.

3. To emphasize the thrust of the original Hebrew language, I have replaced the word "skies" in the 2011 NIV with "firmament."

4. To emphasize human accountability, I have replaced the word "people" in the 2011 NIV with "men and women."

5. M. K. Sateesh, *Bioethics and Biosafety* (New Delhi, India: I. K. International, 2008), 17. I am grateful to geneticist Darrel Falk for his assistance in understanding this scientific information.

6. In previous publications I used "spiritual concordism," but was never comfortable with this term.

Chapter 4: Intelligent Design and the Book of God's Works

1. Michael J. Behe, *Darwin's Black Box* (New York: Free Press, 1996), 39.

2. Ibid., 69–73. Redrawn by Kenneth Kully from D. Voet and J. G. Voet, *Biochemistry*, 2nd ed. (New York: Wiley, 1995), 1259.

3. Tim Wong, et al., "Evolution of the Bacterial Flagellum," *Microbe* 2:7 (2007): 335–40. I am grateful to Ken Miller for this citation.

4. To emphasize the thrust of the original Hebrew language, I have replaced the word "skies" with "firmament" in verse 1 of this 2011 NIV quotation.

5. To emphasize human accountability, I have replaced the word "people" in the 2011 NIV with "men and women."

6. This is commonly known as the "argument from design." But to be more precise, it is an argument from evidence in nature to defend the belief in design and the existence of God.

7. Richard Dawkins, *The Blind Watchmaker* (London: Penguin, [1986] 1991), xiii, xv, xvi, my italics.

8. Also see Richard Dawkins, *The God Delusion* (New York: Houghton Mifflin, 2006), 2, 156–57.

9. Ibid., 31.

10. Paul W. C. Davies, *God and the New Physics* (London: Penguin, 1983), 179.

11. Roger Penrose, *The Emperor's New Mind* (Oxford, UK: Oxford University Press, 1989), 343–44. I am grateful to physicist Don Robinson for introducing me to this book and explaining the mathematics.

12. Peter D. Ward and Donald Brownlee, *Rare Earth* (New York: Copernicus Springer-Verlag, 2000), 16, 275.

13. The list of "right" features in the next two paragraphs comes from Ward and Brownlee, *Rare Earth*, xxvii–xxviii.

14. Simon Conway Morris, *Life's Solution* (Cambridge, UK: Cambridge University Press, 2003), 233–34.
15. Ibid., xii, 151–73.
16. Ibid., 457–61.
17. Ibid., 328.

Chapter 5: Ancient Science and the Book of God's Words

1. John H. Walton, *The Lost World of Genesis One: Ancient Cosmology and the Origins Debate* (Downers Grove, IL: IVP Academic, 2009), 9.
2. Diagram redrawn from Jeremy Black and Anthony Green, *Gods, Demons, and Symbols of Ancient Mesopotamia: An Illustrated Dictionary* (Austin, TX: University of Texas, 1992), 53.
3. Diagrams redrawn by Zondervan from Othmar Keel, *The Symbolism of the Biblical World* (New York: Seabury, 1978), 36, 174.
4. Aeschylus, *Aeschylus I: Oresteia*, trans. and intro. Richard Lattimore (Chicago: University of Chicago Press, 1953), 158.
5. Nicolas Hartsoeker (1656–1725), image from his *Essai de Dioptrique* (1694), image taken from Google Images.
6. See David A. Leeming, *Creation Myths of the World: An Encyclopedia*, 2 vols., 2nd ed. (Santa Barbara, CA: ABC-CLIO, LCC, 2010).
7. Richard J. Clifford, *Creation Accounts in the Ancient Near East and in the Bible* (Washington, DC: Catholic Biblical Association, 1994), 30. This account is also known as "KAR 4."
8. Ibid.
9. Walter Beyerlin, ed., *Near Eastern Religious Texts Relating to the Old Testament* (Philadelphia: Westminster Press, 1978), 75.
10. Clifford, *Creation*, 74.
11. Ibid., 75.
12. Ibid., 48–49.
13. Ibid., 105, 107.

14. David Frost, *Billy Graham: Personal Thoughts of a Public Man. 30 Years of Conversations with David Frost* (Colorado Springs, CO: Chariot Victor, 1997), 73–74, my italics.

Chapter 6: Moving Beyond the "Evolution" vs. "Creation" Debate

1. In response to this debate, I presented a TED talk entitled "Beyond the Bill Nye vs. Ken Ham Debate." It can be found at www.youtube.com/watch?v=QaeGfV-N2kM.

2. "Six in Ten Take Bible Stories Literally, But Don't Blame Jews for Death of Jesus" (February 15, 2004), *ABC News*, www.abcnews.go.com/images/pdf/947a1ViewsoftheBible.pdf.

3. Martin Luther, *Luther's Works, Lectures on Genesis: Chapters 1–5*, ed. Jaroslav Pelikan (St. Louis, MO: Concordia, 1958), 5.

4. Ibid., 3.

5. I will address this issue with Matthew 19 in the final section of this chapter.

6. "Public Praises Science; Scientists Fault Public, Media," Pew Research Center (July 9, 2009), 68, www.people-press.org/files/legacy-pdf/528.pdf (accessed March 7, 2015).

7. This view of divine action was presented earlier in chapter 4 on pages 64–65.

8. Colorado Springs, CO: NavPress Publishing Group, 2001.

9. This analogy was previously introduced in chapter 2 on page 41 and in chapter 3 on page 53.

10. See 1 John 1:1–3; 2 Peter 1:16–18; Luke 1:1–4; Acts 1:1–9. Also see Richard Bauckham, *Jesus and the Eyewitnesses: The Gospels As Eyewitness Testimony* (Grand Rapids: Eerdmans, 2006).

11. New York: Free Press, 2006.

12. Edward J. Larson and Larry Witham, "Scientists Are Still Keeping the Faith," *Nature* 386 (April 3, 1997): 435.

13. See "Secular/Nonreligious/Agnostic/Atheist" in "Major Religions of the World Ranked by Number of Adherents,"

accessed June 4, 2015, www.adherents.com/Religions_by_
Adherents.html.

14. "Not All Nonbelievers Call Themselves Atheists," Pew Research
Center (April 2, 2009), accessed June 4, 2015, www.pewform.
org/2009/04/02/not-all-nonbelievers-call-themselves-atheists.

15. See www.ualberta.ca/~dlamoure/dawkins_and_lamoureux
.html.

─────── Chapter 7: Galileo and God's Two Books ───────

1. John Paul II, "Lessons of the Galileo Case," *Origins: CNS
Documentary Service* 22:22 (November 12, 1992), 371–73.

2. Jeffrey B. Russell, *Inventing the Flat Earth: Columbus and Modern
Historians* (Westport, CT: Praeger, 1991), 62, 69–70.

3. I am thankful to Mark Kalthoff for introducing me to this letter.

4. Galileo Galilei, "Letter to the Grand Duchess Christina" (1615)
in M. A. Finocchiaro, ed. and trans., *The Galileo Affair: A
Document History* (Berkeley, CA: University of California Press,
1989), 93.

5. Ibid.

6. Ibid., 104.

7. Ibid., 93.

8. Ibid.

9. Charles R. Darwin, *On the Origin of Species: By Means of Natural
Selection or the Preservation of Favoured Races in the Struggle for
Life. A Facsimile of the First Edition*, introduction by Ernst Mayr
(Cambridge, MA: Harvard University Press, [1859] 1964), 488.

10. Galileo, "Christina," 105.

11. Ibid., 94, 96.

12. Galileo Galilei, *The Assayer* (1623) in Stillman Drake, trans.,
Discoveries and Opinions of Galileo (Garden City, NY: Doubleday
Anchor, 1957), 237–38, my italics.

13. Ibid., 238.

14. Galileo, "Christina," 93.

15. Ibid., 104.

16. Galileo Galilei, "Letter to Castelli" (1613), in Finocchiaro, *Galileo Affair*, 49.
17. Ibid., 51.
18. Galileo, "Christina," 92.
19. Ibid., 96.
20. Galileo, "Castelli," 51; Galileo, "Christina," 93.
21. Galileo, "Castelli," 51–52.
22. Galileo, "Christina," 94, 106.
23. Ibid., 93, my italics.
24. Ibid., 92, my italics.
25. Ibid.
26. Ibid., 106.
27. Ibid., 92.
28. Galileo, "Castelli," 49.
29. "Six in Ten Take Bible Stories Literally, But Don't Blame Jews for Death of Jesus" (February 15, 2004), ABC News, www.abcnews.go.com/images/pdf/947a1ViewsoftheBible.pdf.
30. Galileo, "Christina," 87.
31. Ibid., 100–101
32. Ibid., 90.
33. Henry M. Morris, *Troubled Waters of Evolution* (San Diego: Creation-Life, 1982), 75.
34. Galileo, "Christina," 93.
35. Ibid.
36. Ibid., 109.
37. Ibid., 96.

Chapter 8: The Religious Evolution of Darwin

1. Richard Dawkins, *River Out of Eden* (New York: BasicBooks, 1995), 133.
2. Richard Dawkins, *The Blind Watchmaker* (London: Penguin, [1986] 1991), 6.
3. Charles Darwin, *The Autobiography of Charles Darwin, 1809–1882*, ed. Nora Barlow (London: Collins, [1876] 1958), 87.

4. Ibid., 49.

5. Ibid., 57.

6. Ibid., 85.

7. R. D. Keynes, ed., *Charles Darwin's Beagle Diary* (Cambridge, UK: Cambridge University Press, 2001), 403.

8. Ibid.

9. Ibid., 444.

10. See pages 68–71 in chapter 4.

11. Darwin, *Autobiography*, 91. See full passage on page 166.

12. Ibid.

13. Ibid., 85.

14. Ibid., my italics.

15. Ibid., 86.

16. See page 120 and endnote 10 in Chapter 6 on page 192.

17. Charles Darwin, *On the Origin of Species: By Means of Natural Selection or the Preservation of Favoured Races in the Struggle for Life. A Facsimile of the First Edition*, introduction by Ernst Mayr. (Cambridge, MA: Harvard University Press, [1859] 1964), 186, 188, 189, 413, 435, 488.

18. Darwin, *Autobiography*, 93. See the context of this quotation in passages below on page 167 and associated with endnotes 38 and 39.

19. Darwin, *The Origin of Species*, 488.

20. Charles Darwin, *Darwin's Natural Selection, Being the Second of His Big Species Book Written from 1856 to 1858*, ed. R. C. Stauffer (London, UK: Cambridge University Press, 1975), 224, also at www.darwin-online.org.uk/content/frameset?pageseq=1&itemID=F1583&viewtype=text.

21. Darwin, *The Origin of Species*, 490.

22. Darwin to Gray, May 22, 1860, Darwin Correspondence Project, Letter 2814 at www.darwinproject.ac.uk/entry-2814; also in Francis Darwin, ed., *The Life and Letters of Charles Darwin*, 3 vols. (London, UK: John Murray, 1887), II:311–12.

23. Richard Dawkins, *A Devil's Chaplain: Reflections on Hope, Lies, Science, and Love* (New York: Houghton Mifflin, 2003), 8.

24. Darwin to Gray, May 22, 1860.

25. Ibid.

26. Ibid., my italics.

27. Ibid., my italics.

28. Charles Darwin, Notebook C [Transmutation of Species (Feb 1838 to Jul 1938)], 196–97; The Complete Works of Charles Darwin Online, www.darwin-online.org.uk/content/frameset?itemID=CUL-DAR122.-&viewtype=text&pageseq=1.

29. Darwin, *The Origin of Species*, 488.

30. Charles Darwin, *The Descent of Man, and Selection in Relation to Sex*, 2nd ed. (London: John Murray, [1871] 1874), 613.

31. Darwin, *Autobiography*, 94.

32. Ibid., 90.

33. Ibid., 88–90, my italics.

34. Ibid., 90, my italics. Also see Darwin, *Beagle Diary*, 444.

35. Ibid., 91, my italics. Also see Darwin, *Beagle Diary*, 444.

36. Ibid., 91.

37. Ibid., 92–93, my italics.

38. Ibid., 93.

39. Ibid.

40. Darwin to J. Fordyce, May 7, 1879, Darwin Correspondence Project, Letter 12041 at www.darwinproject.ac.uk/entry-12041.

41. Ibid.

42. Ibid., my italics.

43. As quoted in Francis Darwin, *Life and Letters*, I:316, my italics.

Chapter 9: Let the Students Speak!

1. Two accessible introductions to this genetic evidence are Francis S. Collins, *The Language of God: A Scientist Presents Evidence for Belief* (New York: Free Press, 2006) and Daniel J. Fairbanks, *Relics of Eden: The Powerful Evidence of Evolution in Human DNA* (Amherst, NY: Prometheus, 2010).